Mute Vol 2 #10

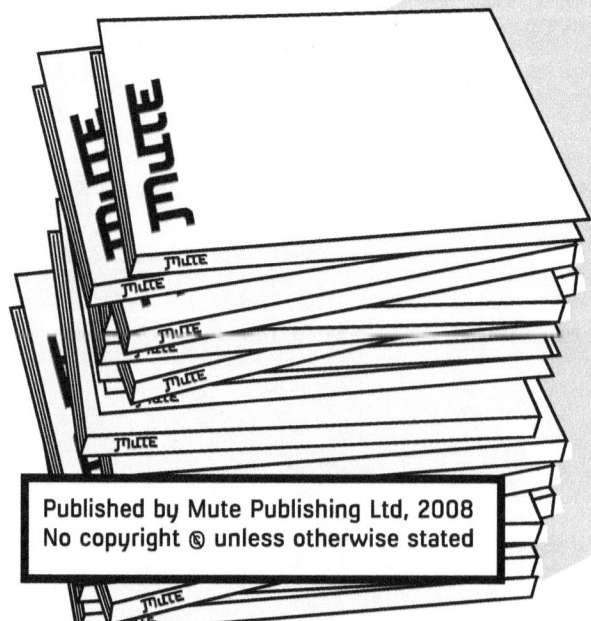

# MUTE Vol 2 #10
## AUTUMN ISSUE – OCTOBER '08

**EDITOR**
Josephine Berry Slater <josie@metamute.org>

**DEPUTY EDITOR**
Benedict Seymour <ben@metamute.org>

**ASSISTANT EDITOR**
Anthony Iles <anthony@metamute.org>

**EDITORIAL BOARD**
Josephine Berry Slater, Matthew Hyland <infuriant@autistici.org>, Anthony Iles, Demetra Kotouza <demetra@inventati.org>, Hari Kunzru <hari@metamute.org>, Pauline van Mourik Broekman, Benedict Seymour and Simon Worthington

**MUTE PUBLISHING ADVISORY BOARD**
Ceri Hand, Sally Jane Norman, Sukhdev Sandhu and Andy Wilson

**PUBLISHERS**
Pauline van Mourik Broekman <pauline@metamute.org>
Simon Worthington <simon@metamute.org>

**ISSUE DESIGN**
Laura Oldenbourg <laura@metamute.org>

**ADVERTISING & MARKETING**
Lois Olmstead <lois@metamute.org>
T: +44 (0)7791284039

**WEBSITE**
Metamute.org is powered by Drupal and CiviCRM FLOSS Software, with additional software services by our very own OpenMute
http://openmute.org

**TECH SUPPORT**
Web infrastructure: Darron Broad <darron@kewl.org>

**INTERNS**
Charlotte Levins
Special thanks to: Hannah Marshall & Simon Morgan – you saved our ass!

**OFFICE**
Mute, Unit 9, The Whitechapel Centre, 85 Myrdle Street,
London E1 1HQ, UK
T: +44 (0)20 7377 6949
F: +44 (0)20 7377 9520
email: <mute@metamute.org>

**SUBSCRIPTIONS**
T: +44 (0)20 7377 6949
F: +44 (0)20 7377 9520
email: <subs@metamute.org>
web: http://www.metamute.org/subs/

**DISTRIBUTION UK**
Central Books,
99 Wallis Road,
London, E4 5LN
T: +44 (0)20 8986 4854
F: +44 (0)20 8533 5821

**DISTRIBUTION NORTH AMERICA**
Please contact:
Lois Olmstead <lois@metamute.org>
T: +44 (0)7791284039
or visit http://www.moreismore.net

**CONTRIBUTING**
Mute welcomes contributions of all kinds.
Email <mute@metamute.org> with your ideas

You can also publish on Mute's website [http://www.metamute.org]. Post news, texts, events and comments, or upload media to the Mute Public Library http://pl.metamute.org

The views expressed in Mute and Metamute are not necessarily those of the publishers or service providers

Mute is published in the UK by Mute Publishing Ltd. and printed by OpenMute
http://openmute.org print on demand (POD) book services in the USA and UK

**COVER**
Pauline van Mourik Broekman

ISSN 1356-7748-10
ISBN 978-1-906496-21-0

Mute is supported by
Arts Council England

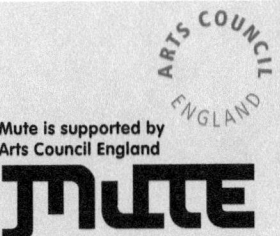

# CONTENTS

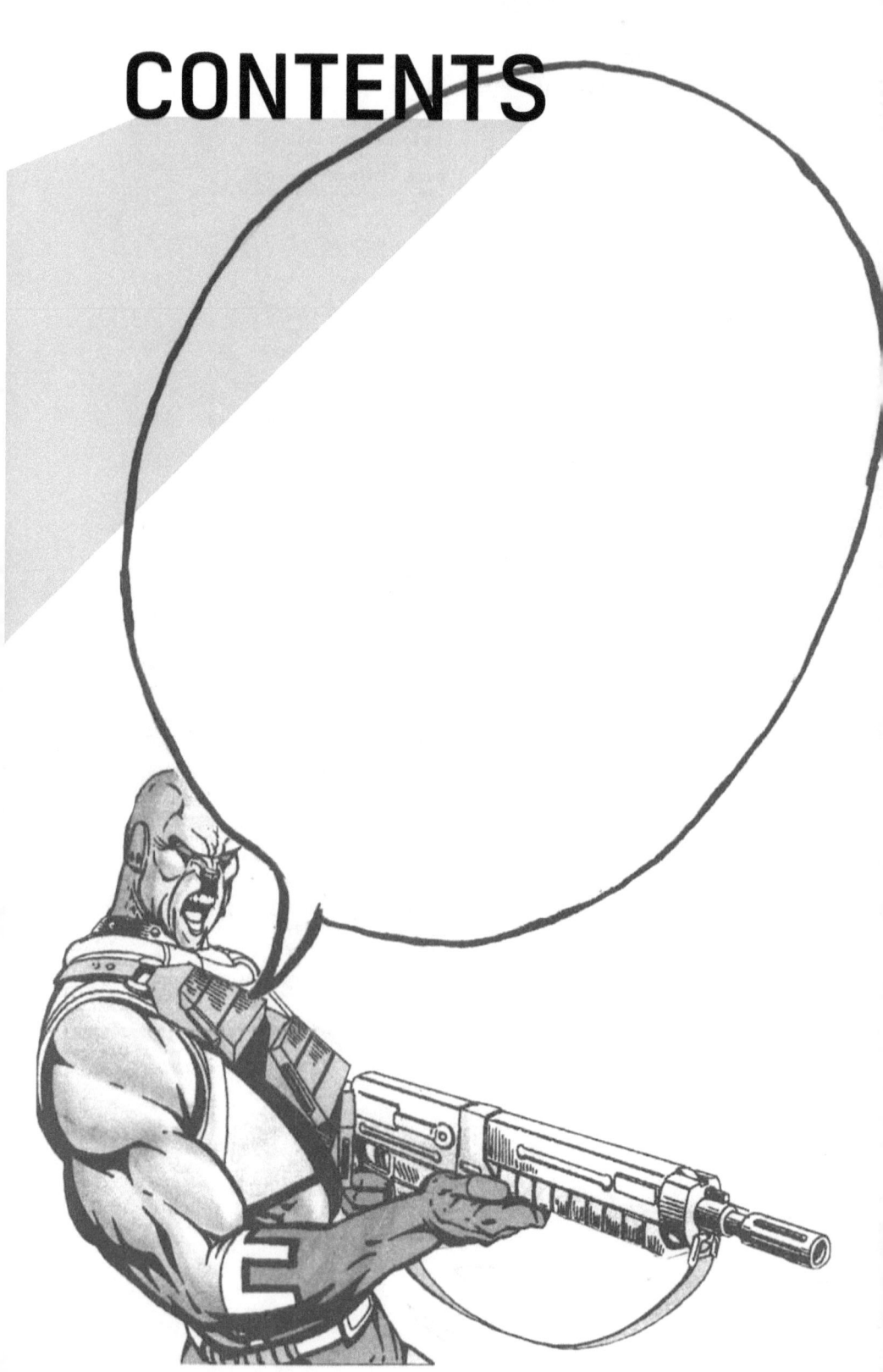

**6 EDITORIAL**
by Josephine Berry Slater

**14 LIVERPOOL – CULTURE OF CAPITAL**
Leo Singer and Clara Paillard crash Liverpool's regeneration party

**24 DESCRAMBLING THE 'FOOD CRISIS'**
George Caffentzis puts the class politics of hunger back on the table

**32 FROM SUBPRIME TO SLUMP?**
Jon Amsden argues that the capitalist doom doctors have got things the wrong way round

**42 MR SMITH GOES TO BEIJING**
Daniel Berchenko corrects Giovanni Arrighi's retrogressive vision of China as the future of capitalism

**54 MEXICAN WAVE**
Mihalis Mentinis foresees a new cycle of armed anti-capitalist struggle in Mexico

**64 ANY OTHER BUT OUR SELVES**
J.J. Charlesworth attacks a curatorial cult of the Other

**78 ORIENTALISM INVERTED: THE RISE OF 'HINDU NATION'**
Neil Gray retraces the rise of Indianness as German ideology, from colonial mystique to neoliberal pogroms

**102 ONE WORLD, ONE LIE**
Paula Cerni finds a thoroughly modern lack of democracy in Tibet

# EDITORIAL

Neoliberal governments want to have their cake and eat it. Having dismantled the social buffer of regulation, public assets and benefits in the name of economic dynamism and personal responsibility, they are now bailing out the financial sector, absolving corporate 'individuals' of responsibility for their past recklessness. At the macro level, we have seen US, UK and European central banks flooding markets with liquidity in a desperate attempt to restart bank lending. Latterly, we have seen tokenistic gestures towards the 'little man' such as the UK government's decision to subsidise the housing market through stamp duty holidays and helping mortgage defaulters. It doesn't take a genius to see that the well-being of business is the top priority of governments.

But what about individuals? In Britain, Thatcher's sell off of social housing was perhaps the most decisive act in throwing the working class on the mercy of the market and credit lenders while appealing to their 'family values' and dreams of personal security. Today the housing market is in deep crisis and the family values card is again being played in the service of the market. The house-buyer's aid package will only encourage more people into a perilous housing market and negative equity, prioritising banks over people's needs. Instead of creating adequate stocks of social housing for the increasing numbers the crisis will make homeless, the money is intended to prop up the housing bubble and the debt-dependent economy it drives. £100 million will be used to help with mortgage interest repayments. Banks must be loving it, as they hike up fixed-rate mortgages to an 11-year high.

Hazel Blears, Labour's community secretary, gets half way to the truth of the situation. The plans to help people, she explains, are not aimed at those who have been reckless in their borrowing but at 'decent' families. That's right! People shouldn't be branded reckless and punished for their heavy dependence on credit when the wage has declined steadily in real terms since the 1970s. The indecent parties are, of course, the banks, hedge funds, private equity companies etc., profiteering from public subsidies. And any notion that this widespread 'recklessness' will be lessened by the 'worst recession in 60 years' is rebutted by a recent report showing that people's dependence on credit has increased over the last year in the UK. Clearly, then, the cost of borrowing has simply become more expensive post-sub-prime, and its terms revised in favour of lenders, while its economic necessity remains unchanged.

So what are 'decent' families to do in such turbulent times? The government itself has predicted that while crime (or is that criminalisation?) will rise in response to lay-offs and general economic misery, right-wing extremist political activity is also set for an upturn. As if anticipating all this, Hollywood offers us the embattled figure of the Dark Knight as the citizen's last hope. But even this heavily armoured bat struggles in the face of nihilistic chaos. Amitabh Kumar's *Raj Comics for the Hard Headed*, which illustrates Neil Gray's article on Hindu cultural nationalism in this issue of *Mute*, considers the complexity of the superhero's popularity. Does he/she represent a God-like power from another dimension, or a manifestation of our human potential to overthrow our oppressors? Does he/she embody the law as state of exception, or act outside the (perennially suspended) law? Whatever the answer, Neil Gray's discussion of Hindu nationalism's racist myth of Indianness shows what can happen when political movements manipulate popular fantasies of identity and salvation. In the case of Hindutva, it is Lord Rama – Vishnu's avatar on earth – who has been transformed into the figure-head of its racist, nationalist politics.

Against the reactionary figure of the God-like individual redeemer, some of the writers in this issue pose the figure of collective struggle. George Caffentzis analyses the links between the food, fuel and housing bubbles. The fight against neoliberalism's 'new enclosures', he argues, is largely responsible for these bubbles, which are essentially capitalism's response to continued resistance. While its success can be measured in terms of the financial punishment it unleashes we can, argues Caffentzis, still read these crises as signs of working class and peasant defiance. Mihalis Mentinis, in his analysis of the growing anti-capitalist consensus across Mexico's social movements, sees violent confrontation on the horizon.

The lone crusader is not the only superhero fantasy doing the rounds. Heroically labouring populations whose toil might somehow sustain the world indefinitely are also popular, even on the left. Giovanni Arrighi's account of China's 'industrious revolution' presents just such a fantasy. However, Daniel Berchenko argues its continued economic growth cannot be sustained by the hyper-exploitation of its workers alone. While China fails to invest in fixed capital and the education and reproduction of its workforce, how sharp is its competitive edge? Such Marxian fundamentals, it seems, don't interest US financial Doomsayers any more than they do sociologists. Here, Jon Amsden argues that the Doctor Dooms are not doomy enough, their predictions of recession all too easily allayed by the Federal Reserve's series of interest rate cuts. In the absence of the promised economic kick-start, the real threat is not inflation but deflation, as demand fails and the real economy contracts.

Between these dark predictions and the State's failure to put in place a safety net for those falling off the credit tightrope, can we expect anything short of increased austerity for the working class? Unless, of course, we reject the superhero fantasy altogether and see that it is ordinary people who have the power to rid Gotham of its ghouls.

---

Josephine Berry Slater <josie@metamute.org> is Editor of Mute

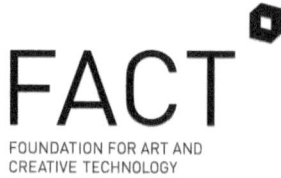

## FOUNDATION FOR ART AND CREATIVE TECHNOLOGY

Until 30 November

## Liverpool Biennial: MADE UP

U-Ram Choe / Ulf Langheinrich / Terrence Handscomb / Michael Bell-Smith / Stella Brennan / Lisa Reihana / Muchen

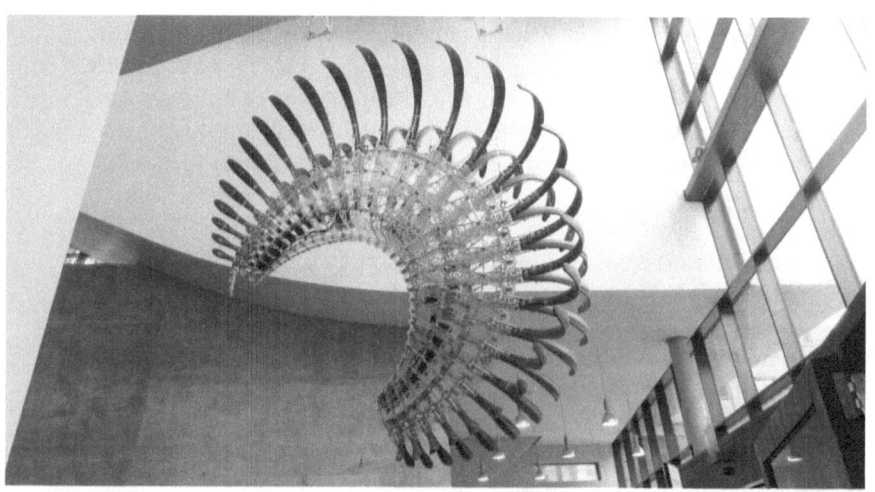

12 December - 22 February

## DING>>D0NG

Andy McCluskey / Peter Saville / Hambi Haralambous / The Fragmented Orchestra / Tetsuaki Baba / Julian Oliver / Pix / The Owl Project / Philip Jeck / Luke Jerram

Explore possible futures of music with instruments, games, gigs, micro-residencies, talks and late-night sleepovers...

**www.fact.co.uk**
88 Wood Street, Liverpool L1 4DQ

Four new titles in Berg's acclaimed series

# THE KEY CONCEPTS

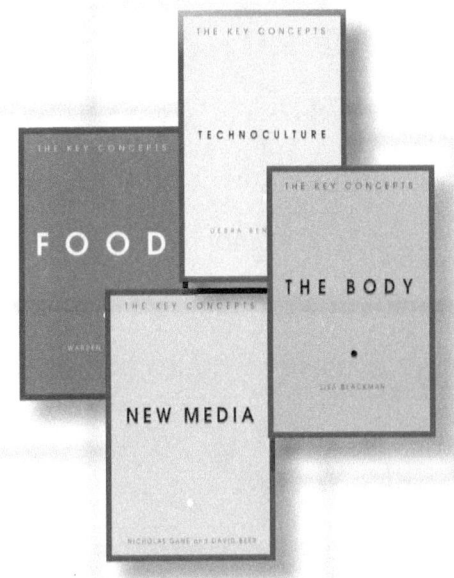

*The Key Concepts* series of texts explores a range of issues in the social sciences and humanities.

Each text covers the core ideas and key cross-disciplinary arguments surrounding the field, mapping out the theoretical terrain across a specific subject or idea.

Order direct from our website at bergpublishers.com and receive **20%** off the recommended retail price

## NOW AVAILABLE

|   | Title | Author | PB ISBN | NOW ONLY * |
|---|---|---|---|---|
|   | Film | Nitzan Ben-Shaul | 9781845203665 | PB £10.40 |
|   | Globalization | Thomas Hylland Eriksen | 9781845205249 | PB £10.40 |
| NEW! | Technoculture | Debra Benita Shaw | 9781845202989 | PB £11.99 |
| NEW! | The Body | Lisa Blackman | 9781845205904 | PB £11.99 |
| NEW! | Food | Warren Belasco | 9781845206734 | PB £11.99 |
| NEW! | New Media | Nicholas Gane & David Beer | 9781845201333 | PB £11.99 |

* prices quoted are the discounted rates for orders placed via the website

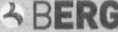

to order visit bergpublishers.com

# Mediamatic

## Any Media Documentary workshop @ IDFA
## November 21 - November 26, 2008

Mediamatic and IDFA organize a 5-day workshop where you will explore a range of different media tools to augment your documentary practice. A special focus of this workshop will be creative information visualisation, using other tools than the camera to make documentary projects. All participants will make a working prototype of their AnyMedia documentary project.

*www.mediamatic.net/workshops*

---

 **WALLFLOWER PRESS | INDEPENDENT SPECIALIST PUBLISHING IN CINEMA AND THE MOVING IMAGE**

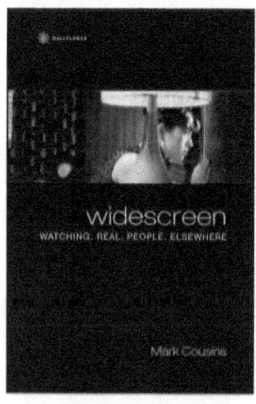

## widescreen
### WATCHING. REAL. PEOPLE. ELSEWHERE
Mark Cousins

£14.99 pbk 978-1-905674-62-6
£45.00 hbk 978-1-905674-79-3

Cinema has undergone huge changes in the last decade: Asian filmmaking has been making the running; the ne'er do well genre, documentary, has broken through; digitisation and DVD has revived film history and is revolutionising projection; world cinema has shifted in the direction of the real and the visually grainy; and animation has become more dominant than at any time since Disney. Month by month, in the acclaimed journal *Prospect*, critic and filmmaker Mark Cousins has charted and contextualised these changes. Writing from Britain, Europe, Iran, India and Mexico, he has looked at the social trends and aesthetic implications of modern cinema's shifting sands. *Widescreen: Watching. Real. People. Elsewhere* is the result; a sceptical, passionate, eye-witness account of film today, argued originally and written with panache.

**for further information on Wallflower Press titles, and online discounts of up to 20%, please visit www.wallflowerpress.co.uk**

---

WALLFLOWER PRESS | 6 Market Place, London W1W 8AF | tel: 020 7436 9494 | email: info@wallflowerpress.co.uk

# The MIT Press
http://mitpress.mit.edu

Fitzroy House, 11 Chenies Street, London WC1E 7EY
tel: 020 7306 0603 • orders: 01243 779 777

## The Privacy Advocates
### Resisting the Spread of Surveillance
#### COLIN J. BENNETT

This book analyses the people and groups around the world who have risen to challenge the increasingly intrusive surveillance practices used by both government and corporations to capture, process and disseminate personal information. Drawing on extensive interviews with key individuals in the movement, Colin Bennett describes a network of self-identified privacy advocates who have emerged from civil society — without official sanction and with few resources, but surprisingly influential.

£18.95 • cloth • 300 pp. (11 illus.) • 978-0-262-02638-3

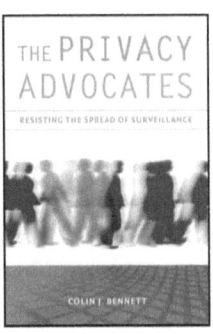

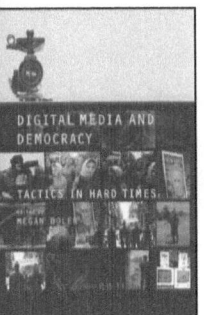

## Digital Media and Democracy
### Tactics in Hard Times
#### EDITED BY MEGAN BOLER

In an age of proliferating media and news sources, who has the power to define reality? *Digital Media and Democracy* explores the contradiction at the heart of the relationship between truth and power today: the fact that the radical democratization of knowledge and plethora of sources and voices made possible by digital media coexist with the blatant falsification of information by political and corporate powers.

£25.95 • cloth • 474 pp. (21 illus.) • 978-0-262-02642-0

# more is more

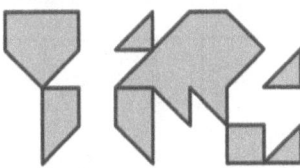

## Independent media distribution & video screening network

More is More is an open source, online distribution system for small-scale and independent media. The aim of the network is to provide independent media producers and cultural organisations with a platform that can connect them to local outlets and events. More is More facilitates the sale of goods at such locations as well as direct through the website itself.

Commercial distributors are not best geared to the distribution of media products from the cultural, non-profit or political sectors. OpenMute's distribution network is an attempt to develop an alternative. More is More distributes the following: video, magazines, books, comics, posters, flyers and music. It is also possible to arrange your own event or film-screening through the platform.

While the site is at an alpha stage, we are looking for reactions and input from individual media-producers, cultural and activist organisations as well as a variety of outlets that might be interested in putting their products online or selling them locally.

**An OpenMute project**

**Supported by Digital Pioneers**

# moreismore.net

# Fourteen years in a nutshell!
## A new book about Mute...

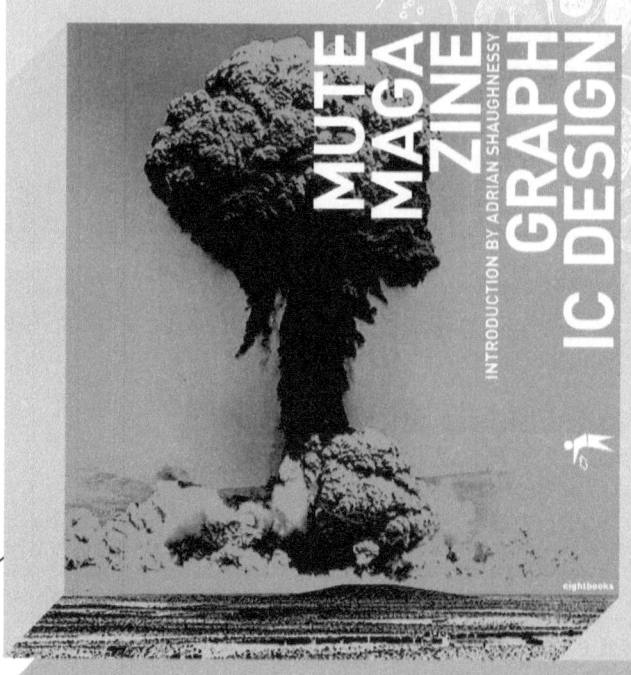

### *Mute Magazine: Graphic Design* (April, 2008)

In the early 1990s, long before the internet became an integral part of life, a handful of pioneering magazines took it upon themselves to imagine the web into existence. Using fiction, interviews, speculative theory and experimental graphic design, London-based Mute wielded an influence disproportionate to its scale. Nearly fifteen years after its launch in November 1994, Mute's publication history defines an era, telling the fascinating tale of one publisher's relationship with the 'digital revolution'. This graphic design history presents a full overview of Mute's output, including logos, covers and spreads.

Introduction by Adrian Shaughnessy, with further contributions from Damian Jaques, Pauline van Mourik Broekman and Simon Worthington.

Published by 8books and now taking orders at Metamute.org/mutegraphics

Softback 220 x 220 mm, 144 pages, 250 colour Illustrations

### 10% discount for Mute subscribers

Buy it online at:
**metamute.org/mutegraphics**

UK £ 19.95       Europe €25
US $ 35          ROW €25

# METAMUTE

## If you *must* have your heroes, try ours out for size at Metamute.org

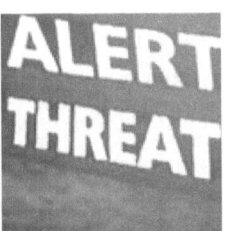

### The Sleep of Realism Produces Monsters
by Andrew Fisher
http://www.metamute.org/en/content/the_sleep_of_realism_produces_monsters
Giving a critical survey of the documentaries of Adam Curtis, Andrew Fisher evaluates the claims to 'realism' and political neutrality made for his work against the critical methodologies of Guy Debord and Georg Lukács

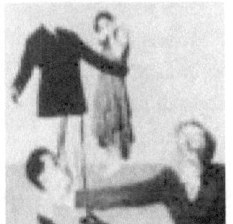

### Burdened by the Absence of Billions
by Howard Slater
http://www.metamute.org/en/content/burdened_by_the_absence_of_the_billions
Marx's concept of 'species being' is for some a way of re-connecting with fertile currents in the communist left. Howard Slater explores Frére Dupont's recent book *Species Being and Other Stories* as a vehicle of exodus from left orthodoxies

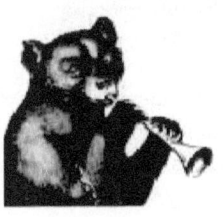

### Great Game II: America Lashes Out on the Borders of China and Russia
by Loren Goldner
http://www.metamute.org/en/content/great_game_ii_america_lashes_out_on_the_borders_of_china_and_russia
The 19th century 'Great Game' rivalry between Britain and Russia for supremacy in Central Asia is seeing a resurgence, with America taking Britain's place. The stakes are higher than ever, argues Loren Goldner

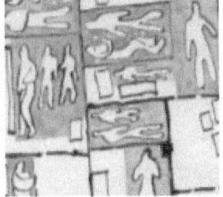

### Monstrous Plans & Good Habitats
by Mark Crinson
http://www.metamute.org/en/content/monstrous_plans_good_habitats
Was modernism complicit with colonialism, and did the struggle for decolonisation also entail a targeting of imperial modernist architecture? Mark Crinson visits the exhibition In the Desert of Modernity to see if the charge will stick

## http://metamute.org

# LIVERPOOL – CULTURE OF CAPITAL

Reporting on the conference Capital, Culture and Power in Liverpool, Leo Singer and Clara Paillard crash the regeneration party and pose some difficult questions for its hosts

*– Stuck in its glare we lose sight of structure*
Roy Coleman speaking of Liverpool's urban patriotism

For three days in July, Liverpool University and Liverpool John Moores University hosted a critical conference about urban regeneration in Liverpool. Both its size and the high proportion of activists present were unusual for an academic conference. Of 155 participants nearly half were non-academics: community activists, community workers, artists and working class activists.

The conference was organised by the European Group for the Study of Deviance and Social Control, an association of radical sociologists and criminologists. In the middle of Liverpool´s year as European Capital of Culture, the organisers decided to devote much of the conference programme to dissenting voices.

The city has recently been submerged under an avalanche of discourse dealing with our everyday lives. The 'creative industries' (including the art house cinema, art galleries, and local media) have been especially active in the field of ideological production. They have been commissioned to deliver various cross-class messages to the city residents. FACT (Foundation for Art and Creative Technology) and Radio BBC Merseyside were asking questions such as: Where are 'we' going to? Who are 'we'? Are 'we' more free than 'we' think? Where are the boundaries of 'our' bodies? What is the identity of this city?, etc. Discourses celebrating the new, post-industrial and cosmopolitan 'community' subtly sweep away the outdated *Boys From The Black Stuff*-like traditions and any sources of identity coming from the working class world.

A number of readily available and servile 'creatives' ensure that the ideological production of the ruling class is sold to traditionally suspicious Scousers in a non-intrusive and 'cool' way. Various dialogical, participatory, experiential platforms, venues and projects have been developed to achieve this task.

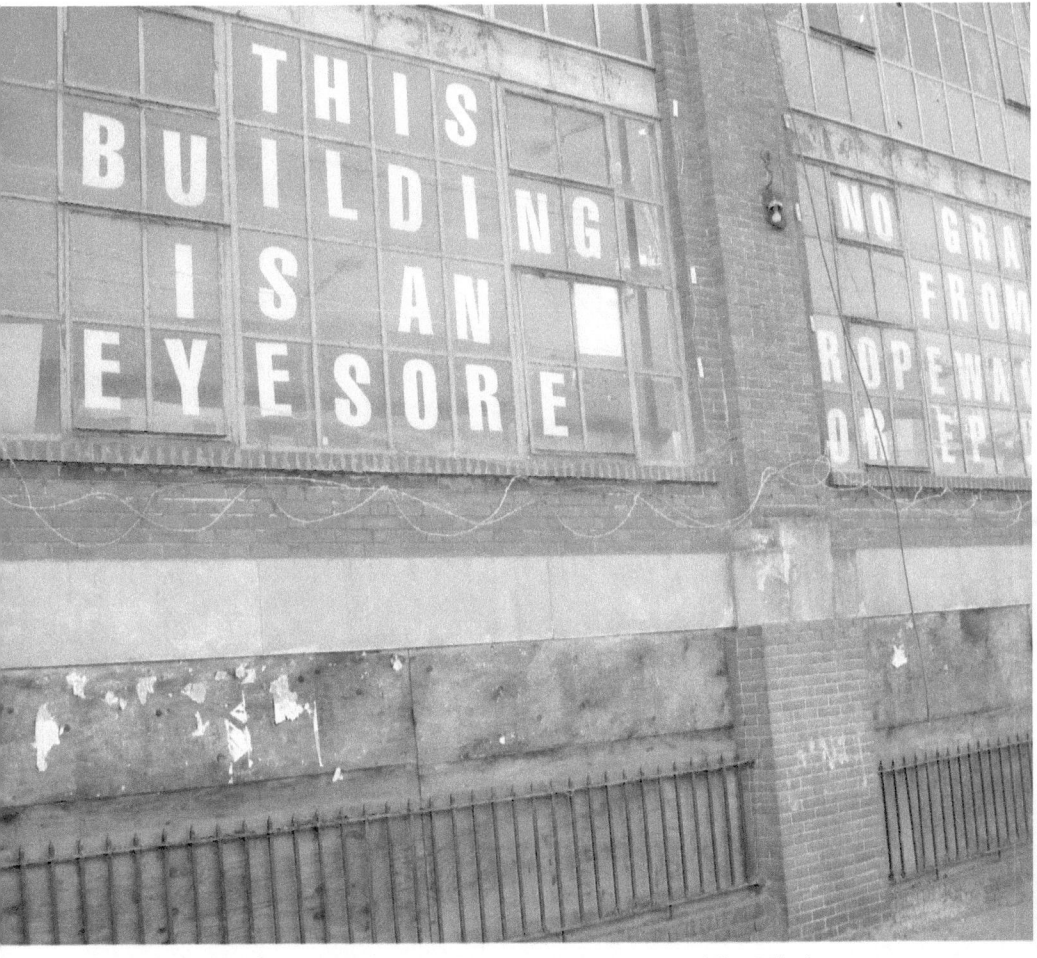

Image: Anti-regeneration in Liverpool's Chinatown. All photographs by Leo Singer and Clara Paillard

## Mutant Patriotism

One example of how cynical inhabitants are turned into participants in the year-long cultural party is the Super Lamb Banana phenomenon. The original work by the Japanese artist Taro Chiezo, commissioned for the Art Transpennine Exhibition in 1998, was conceived as a protest against genetically modified food. The Super Lamb Banana is, as the name suggests, a heroic composite of lamb and banana, a reference to Liverpool's history of exporting lamb and importing bananas. The City of Culture grasped the potential of this piece of public art and turned it into a symbol of Liverpudlian 'wackiness', thus packaging all sorts of stereotypes about Scousers into a much-celebrated commodity. *The Liverpool Echo* is exemplary in whipping up this kind of hysteria. One of the top articles recently reported on a group of local scouts

Liverpool – Culture of Capital

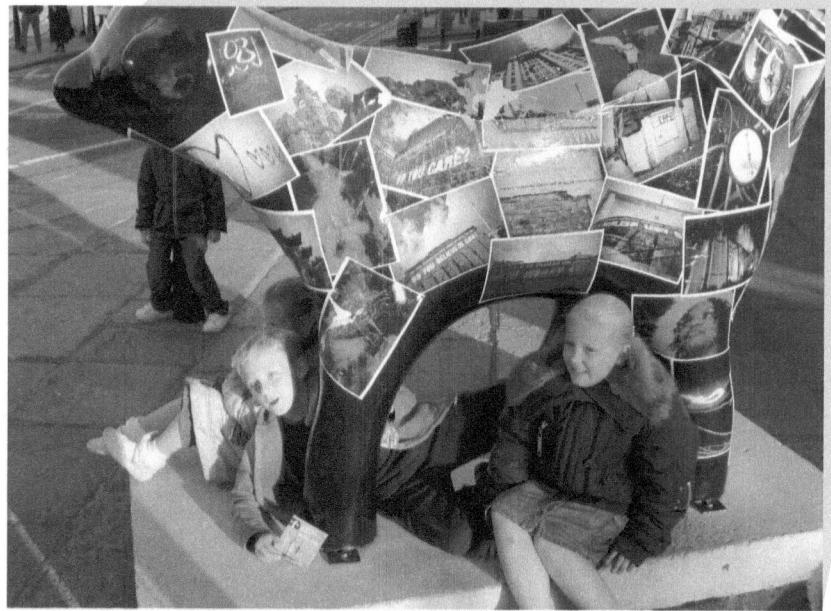

Image: One of the many miniature replicas of Taro Chiezo's *Super Lamb Banana*

who within 24 hours managed to visit all 200-plus Lamb Banana statues scattered over the city! As the scout leader put it:

> We've got photographs of every Superlambanana we visited and we're now trying to register the attempt with the Guinness Book of Records and hope it will be made into an official record.

And by the way, the *Echo*'s article makes sure that every reader gets the message that this Super Lamb Banana hunt was sponsored by the companies Honda, Barclays, BBA Aviation and the city's Comet stores...

Experiencing the Lamb Banana hysteria, one immediately understands the concept of urban patriotism, introduced by Roy Coleman during his keynote speech at the conference. He defined urban patriotism as a strategy of rule for regeneration managers that appears apolitical, banal and funny, thus mystifying class relations. Like all forms of patriotism this kind is narrow-minded, focused on simple images, emotive, celebratory in nature and lacking in reasoning. It is organised around a kind of pride and a 'love' of consumption, 'heritage', iconic buildings or objects like the Lamb Bananas, etc. And like the national variety, urban patriotism glosses over and

Leo Singer and Clara Paillard

deflects attention from the less glamorous aspects of regeneration, including class division, continuing poverty and marginalisation of communities. Coleman continues: 'It involves universities, political decision-

## 35 percent of city centre apartments were vacant even pre-credit crunch

making, marketing, retail sector, policing'. Despite the bombastic and self-confident language employed in this strategy, it is a sign of both the incapacity to face up to reality and 'a means of defence against those deemed to undermine "the brand" or the strategy'.

## Bubble Economics

Private developers and speculators have been playing the regeneration game for years in Liverpool. But it is important not to lose sight of the fact that the bulk of city regeneration has been financed from public sources (EU Objectives 1 and 2, governmental subsidies, New Deal for Communities, etc.) The major private investment is the £1 billion Liverpool One Project, probably the largest single private development in Europe funded by big retail money. This 42.5 acre shopping site in the heart of the city – policed by private security and omnipresent CCTV cameras – was

described as a 'retail republic' in a *Guardian* article from May this year.

Stuart Wilks-Heeg's (University of Liverpool) myth-busting analysis showed that the growth of the city in the last 10 years was not driven by the new economy but rather by the expansion of the public sector (65 percent net job growth). Since 1997, 10,000 manufacturing jobs were lost and 14,000 new jobs in the service economy (call centres) were created. Real unemployment is still around 15-25 percent and the level of skills among workers remains very low. The vulnerability of capital's composition in Liverpool lies in its reliance on public subsidy and cheap credit to sustain the local housing market and construction. So, with 35 percent of city centre apartments vacant even before the credit crunch and the new Liverpool One opening just at the moment that a sharp drop of consumer spending is expected, where is the city heading?

## Cue the Quangos

Liverpool has become the theatre for the worst comedy of the century, with luxury flats germinating like weeds and shopping malls flourishing on any land that's grab-able. But behind the private sector's iconic absurdities hide a myriad of semi-public agencies, populating the local political space at great speed. One after another they usurp the stage of local democracy, pushing out elected representatives and local citizens. This new theatrical

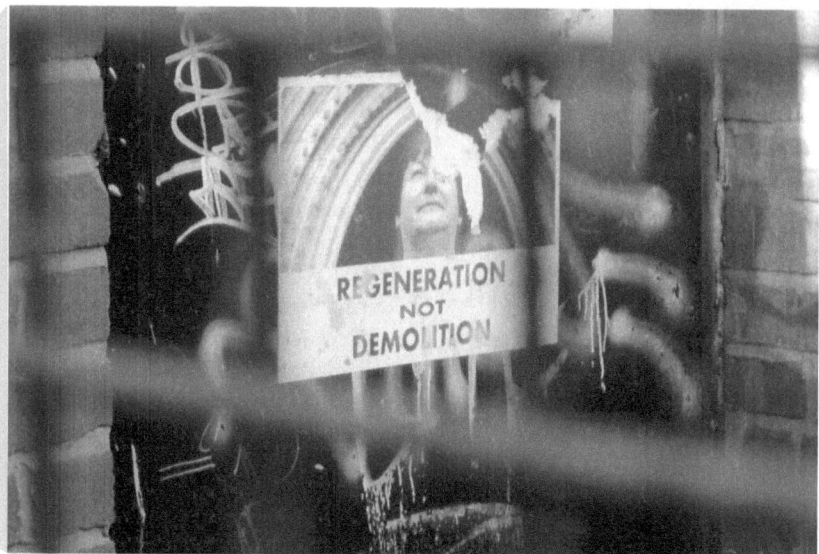

Image: Poster featuring OAP Elizabeth Pascoe who successfully claimed that a compulsory purchase order forcing her to leave her home in Adderley Street breached her human rights, thus delaying an English Partnership project to demolish 500 homes in the area to make way for a widened road into Liverpool city centre

troop are faithful to the Thatcherite script, albeit in a version reanimated by New Labour and its prophets.

The first quango to appear in Liverpool was the Merseyside Development Corporation (MDC), established after the Toxteth uprising in 1981. The MDC was put in charge of regenerating the Albert Dock, the so-called 'jewel in the crown' of Liverpool's 'renaissance'. The Blair era produced new characters to enrich the local 'quangocracy' with the North West Development Agency (NWDA) set up to help create an environment in which businesses in the region could flourish.

## Liverpool relies on public subsidy and cheap credit for housing

This was soon followed by its offspring, Liverpool Vision, an urban regeneration company created to transform the city into one of the UK's leading business destinations by means of an ambitious and far-reaching regeneration programme. Kensington Regeneration, a quango set up to invest in the working class area of Kensington, was funded by the New Deal for Communities. This programme was launched in 1998 as a key part of the Government's strategy for 'regeneration in the

39 most deprived areas across the country'. The package for Liverpool is £62 million. And last but not least the Housing Market Renewal Initiative (HMRI) arrived to deliver the final blow to local inhabitants in other neighbourhoods. Today the stage is crowded with these quangos, disguised as community initiatives and local projects.

But what has been done for the people of Liverpool? Gentrifying the inner city with bulldozers? Redesigning entire communities? Spending millions on 'culture' to allow the high-flyers of the hospitality industry to enter the city? Selling public land to property developers? The North West Development Agency (NWDA) subsidises private profit by handing huge sums of money to speculators. In Kensington, entire Victorian streets are being demolished to allow land assembly to operate in favour of developers.[1] People are being evicted without discrimination from both social and private housing by the New Deal. In the city centre, Liverpool Vision has sold the waterfront to any developers able to build over twelve storeys and has made helping the private sector fill its pockets a priority, just in time for the present 'crisis'.

## Spoiling the Party

Generally, urban regeneration is led by a forced and false consensus rather than public debate, to the point where there is little discernible difference between marketing and public consultation processes. A key issue at the conference was how policy-driven research is incorporating academics, promoting the selective adoption of academic arguments and the misuse of public consultation.

A prime example remains the saga of the 'Fourth Grace'. A 'tall iconic building capable of attracting more visitors to the city and based on Liverpool's culture and history' was supposed to complement the famous Three Graces – industrial skyscrapers erected to celebrate the city's boom a hundred years ago. A public exhibition was held in 2002 to give the public a chance to have their say about the design. Four projects were presented in a catwalk fashion with models and 3D fly-through visualisations. Visitors were asked to 'vote' for their preferred scheme. When Liverpool Vision's decision was announced, the whole city was stunned: they had chosen the least popular building to be built on the waterfront. However, after two years of public spending to promote it, the project collapsed in 2004.

In this concert of demagogy and commodification, activist voices can rarely be heard. Fortunately, they could at the conference. *Nerve* magazine, promoting grassroots culture and organising in Merseyside, hosted a session where residents gave their accounts of the hidden face of the Capital of Culture mega-party.

Liverpool – Culture of Capital

Hazel and Stella from the Granby Triangle testified to the 'lie of consultation' that led to the decision to demolish hundreds of homes in their area:

> All what they could ask was 'What sort of houses do you want?' and we kept saying 'the ones we are in now', they never listened.

Their homes are earmarked for a 'regeneration' scheme that will demolish perfectly good Victorian and Georgian houses to make way for new developments that are unlikely to last more than half a century.

Nina Edge, who lives in the 'Welsh Streets' area in Toxteth, told the story of her community of South Liverpool:

> On one side of the road, the houses are branded 'Victorian 5-bed houses with view on Princes Park' and are valued at £350,000 and on the other side, they are 'derelict dwellings unfit for habitation' and owners are offered £60,000 for them, how is that?

Elisabeth Pascoe, a campaigner from Edge Lane, is still fighting tooth and nail against the Compulsory Purchase Order (CPO) imposed on her house and those of hundreds of other people in her community. The use of CPOs (based on the 1993 Housing and Urban Development Act) means mobilising the State's brutal powers – originally designed to provide for the re-use of brownfield areas – against a poor community. Pascoe warns that this sets an important precedent relevant to residents in housing areas all over the country.

People in the affected communities have been put under stress and troubling rates of mortality have been observed in these areas (33 people in Edge Lane and 9 in the Welsh Streets). Rather than helping local people get a better place to live, the master planners have bulldozed and divided local neighbourhoods. But local people are fighting back and by the end of the conference a support network had been set up bringing together local residents and researchers for the first time.

While it is true that local people have been organising against the demolition of homes, and to protect local heritage and parks, the opposition has been fragmented, often focusing on individual issues rather than linking together. The strong presence of activists and working class people at the conference would probably not have been possible without the effort and authority of the *Nerve*. During a period marked by the relative absence of visible class struggles in this dormant, formerly radical city, the magazine has served as a link between area-based groups, un-organised left-wing individuals, artists and academics. It is a rallying point for people with experience of different political generations and movements: the mass workers' struggles in the

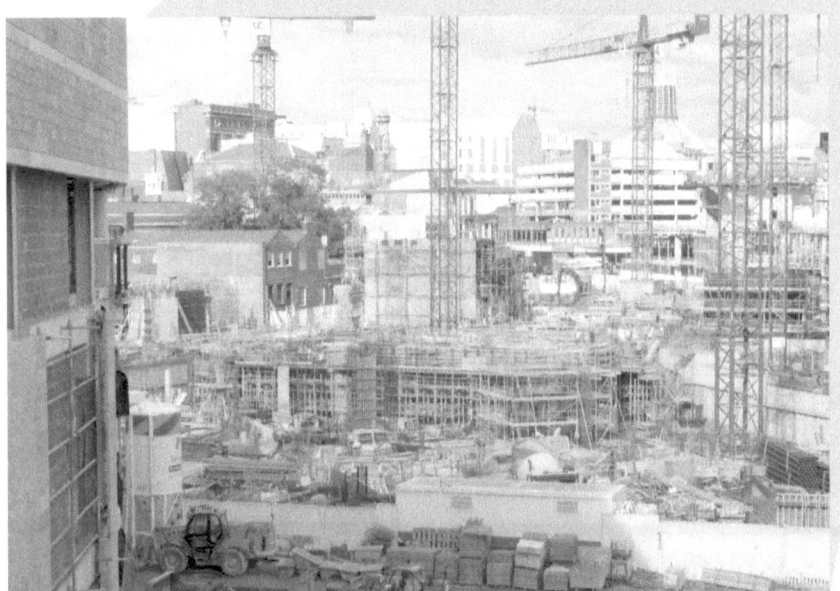

Image: Construction of Liverpool One shopping and leisure district. One of these cranes, the property of Norfolk based Falcon Crane Hire company, collapsed in January 2007, killing a Polish worker, Zbigniew Swirzinski. Since he was self-employed and working without a contract his family hasn't received any compensation. Families Against Corporate Killers (FACK) are demanding a more just investigation. Meanwhile, the same type of crane remains in use

1970s, the unemployed youth and squatters' movement in the 1980s, and a more diverse spectrum of social struggles (including anti-regeneration campaigns) from the 1990s on.

## The Role of the University Revisited

The conference was an important statement against the self-congratulatory City of Culture rhetoric. It is a paradox that a radical conference was hosted by two universities which are key partners in Liverpool Vision as well as owners and developers of important properties in the city centre. The universities serve as an informal think tank for the City of Culture project. Only a few metres from the conference rooms, a team of sociologists commissioned by the City Council and Liverpool Culture Company were working on their research on regeneration. This project, 'Impacts 08', has had little impact on the regeneration debate other than to support its marketing strategy.

We pointed out this split among academics to sociologist David Whyte, one of the conference organisers, who outlined a complex picture of the role of universities today. He argued that they are neither autonomous spaces of open free discussion nor uniform knowledge factories producing analysis for the State or private sector companies. They are rather terrains of constant struggle where the attachment to academic freedom can still provide cover for radical research. For Whyte, academics often exaggerate the oppressiveness of university commodification in order to justify their passivity and pursuit of government grants. Neoliberalism does exert disciplining pressure but university workers still can organise against that:

## Liverpool – Culture of Capital

The fact that critical work emerges from university ground is not a function of the institution itself. It is always a result of struggles that create the space for this work.

Some of the organisers conceived the conference as an opportunity to build up links with working class and social movements. It is questionable to what extent this was achieved. In November 2008 Liverpool's new BT Convention Centre will be the venue for a conference on the city's regeneration entitled 'On the Waterfront'. Sponsored by the city, the quangos and (surprisingly?) Liverpool University, anyone who wants to get into this event will have to pay £250. Of course no potential troublemaker has been invited to this 'intellectual' celebration of policy makers, developers, consultants and other urban patriots. Compared to this event, the Conference on Capital, Culture and Power stands out as a far more democratic enterprise. No entry fees, free food and drinks at a party in the university atrium, and friendly debates regardless of 'who is who'.

On the other hand, the conference 'on Capital' did not really deepen our understanding of how workers, to whom the academics want to relate, produce capital in their everyday lives. Some of the organisers – Gramscian Marxist sociologists – research conditions of work on construction sites in Liverpool city centre and support campaigning families of workers killed at work. But their serious activity can't replace our research into other related questions. What is the class composition of the new industries or 'upgraded' areas? What strategies do call centre, restaurant, or shop workers use to cope with the demands of their bosses or clients? Are there any experiences of or possibilities for struggle here? One can dismiss these questions claiming that 'the conference wasn't about that.' But this is exactly the problem.

New Labour's urban policy project is about delivery of public resources to private projects. To sustain this sophisticated managerial operation the State needs to generate a highly complex web of governance mechanisms. It is exactly here that the plethora of quangos and 'community empowerment' programmes with their own circular languages are located. It seems that most critical academic research follows this moving target, shifting attention away from something more fundamental – the transformation of work in a 'booming' city and how this feeds the drive of capital.

If the academics who participated in the conference want to demonstrate a commitment to the working class, they should spend less time confronting New Labour's urban policies, various local hegemonies and quangos and instead turn toward the everyday processes of the constitution of work. Because, while some of us may become victims of urban regeneration or crime in our neighbourhoods, all of us have to sell our labour power in one form or another.

Leo Singer and Clara Paillard

## Info

Capital, Culture, Power: Criminalisation and Resistance was hosted by the University of Liverpool and Liverpool John Moores University on behalf of the British-Irish Section of the European Group for the Study of Deviance and Social Control, 2-4 July, 2008

## Footnote

1 Land assembly is a term used in planning to explain the activity of planners and/or professionals who seek to bring together small parcels of land (private and/or public) to create a larger parcel. This is an exercise often used to provide more space for large-scale urban projects (public or private). Sometimes the public sector can seek land assembly to build public facilities but at the moment it merely provides land to housing developers to create large private housing developments for maximum profit.

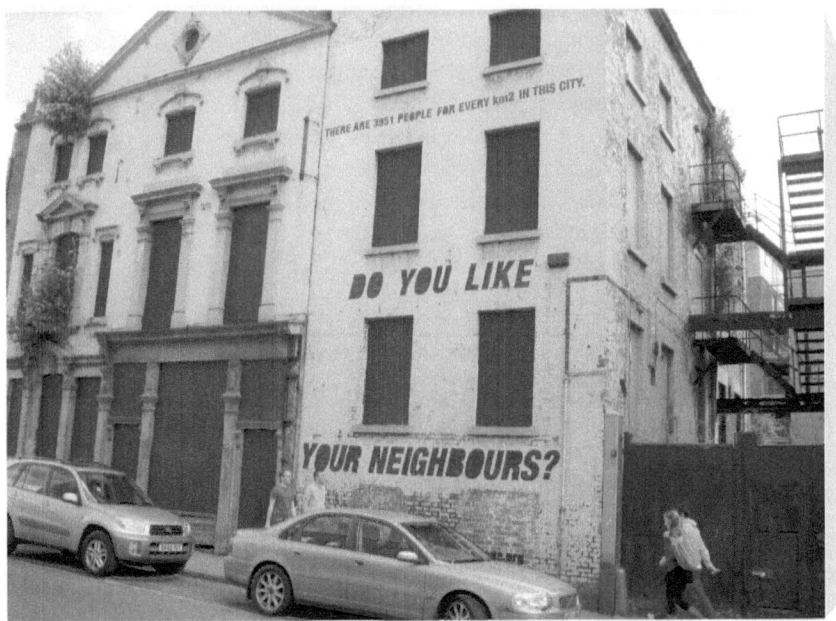

Image: Stockholm artists collective A-APE' s interventions on walls in the city centre. Project commissioned by the Liverpool Culture Company as part of European Capital of Culture 2008 and managed by Liverpool Biennial

Leo Singer is a community worker and a drop in the new East European migrant 'wave' to the UK. Together with Clara Paillard, as part of a larger research group, he is researching the class composition of the hospitality sector in Liverpool's city centre

Clara Paillard <c.paillard@liv.ac.uk> is a French political activist and member of the Campaign for a New Workers' Party. She is completing her PhD on urban regeneration and democracy at University of Liverpool. Clara coordinates an Alternative Programme to the Capital of Culture '08. More info at: www.myspace.com/cityofculture08

# DESCRAMBLING THE FOOD CRISIS

When the world's hedge funds turned from real estate to grain speculation the poor picked up the tab. But the food bubble was no accident, argues George Caffentzis, it was class war

After more than three decades of relative stability, food prices have dramatically increased over the last three years. Between May 2007 and May 2008, corn prices increased by 46 percent, wheat by 80 percent, soybeans by 72 percent; rice by 75 percent. As a result, according to the UN World Food Program, another 130 million people have been added to the hundreds of millions already starving or suffering from malnutrition.

Not surprisingly, then, in dozens of cities across the world, from Port-au-Prince to Cairo to Manila, people have rioted in protest against the economic death sentence imposed on them, clearly aware that fluctuations in commodity prices are not 'facts of nature'. Indeed, the hieroglyphics of food prices both hide and reveal a world of plans, policies and struggles that we need to decipher if we are to explain the roots of this 'food crisis'.

## Perfect Storms and Irrational Bubbles

Our first step must be to reject its characterisation as a 'perfect storm', that is, the outcome of a combination of factors that *nobody* could have possibly predicted.

Nothing could be further from the truth. As with the 'energy crisis' that is sending oil prices to the sky, the rise in the price of staple foods was easily predictable and indeed forecast by analysts and activists across the globe. For years, greens, eco-feminists and, above all, members of peasant movements denounced the neoliberal policies imposed by the World Bank and IMF on the countries of the Global South in the name of economic recovery and structural adjustment. These had disastrous effects on food production and people's ability to reproduce themselves.[1]

Nevertheless, structural adjustment remains to this day the 'Bible' for the regulation of the economies of Africa, Asia, and Latin America. Accordingly, governments in these regions have been pressured to privatise land tenure, cut subsidies to farmers, redirect agricultural production toward export goods, while opening the door to food imports (especially from the US and the EU). In addition, they were urged to dismantle food reserves, the argument being that such protective mechanisms have no place in a free trade, global economy where food is easily and cheaply at hand.

Even when evidence accumulated that these policies created near famine conditions and major social dislocations, making millions of people dependent on the vagaries of the international market, objections were dismissed. 'Experts' appealed to the principle of 'comparative advantage', and deemed it 'scandalous' for countries of the South to demand *food sovereignty*, i.e., 'the right of every population to decide what to eat and how to produce it, having access to land and low interest loans', when these countries' staple foods could presumably be 'more efficiently' produced in the US.[2] As a result, in the space of two decades, many countries that had been totally self-sufficient in food production became net importers of food stuffs from Europe or the US, and millions of small farmers were ruined, forced to migrate to towns or abroad. Many in India, crushed by debts, were driven to suicide (150 thousand in recent years).[3]

In Europe as well, since the formation of the European Common Market, the possibility of food self-sufficiency was undermined, as the objective shaping agricultural production has been the maintenance of a profitable price structure, even though it is achieved through the destruction of much wealth. Building a highly profitable agricultural sector, in fact, has meant instituting quotas regulating what quantity of each product a member country is allowed to produce, and imposing stiff fines on those who exceed these measures. Depending on the country and the particular quotas assigned to it, dairy farmers have been paid to kill their 'surplus' cows, so their milk production would not exceed the limits prescribed, and have been fined when they did not comply, while other farmers have been forced to uproot fruit trees, destroy 'surplus' crops, and so forth.

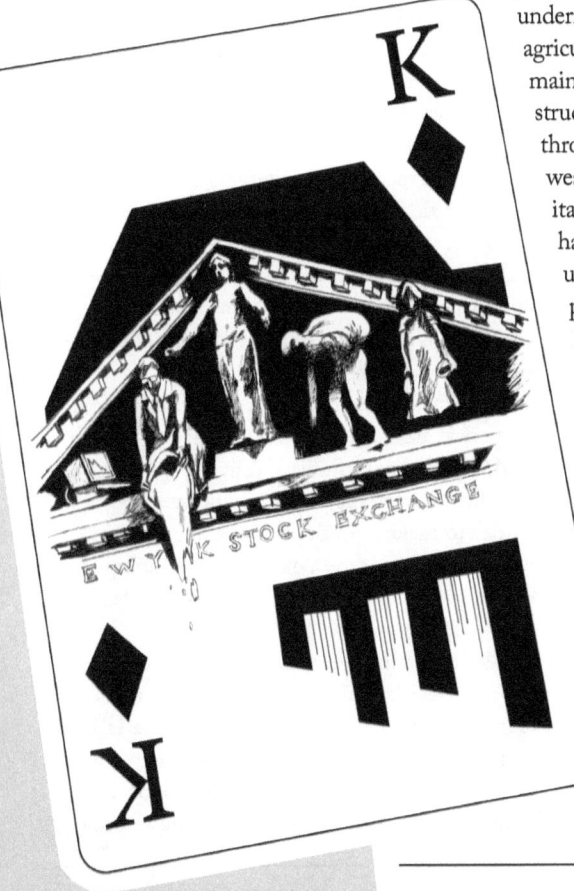
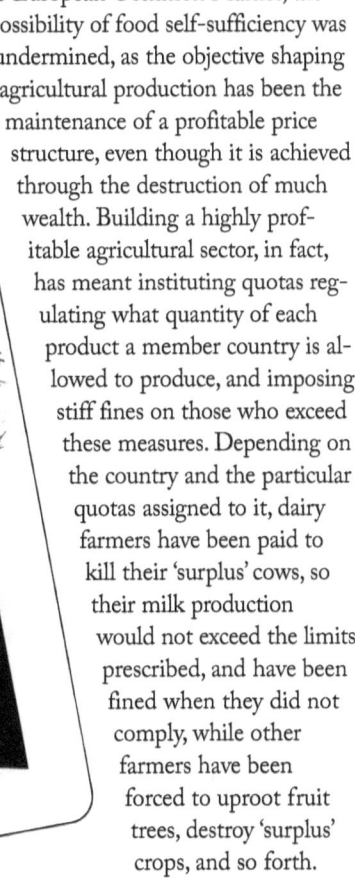

Images: Nick Brooks

In sum, while the official rhetoric espoused by the Food and Agriculture Organization (FAO) has hailed the goal of universal 'food security', the production of artificial 'scarcity' in the service of higher profits and the elimination of subsistence farmers has been the true guideline of international food policy for quite some time.

Under these circumstances, to speak of the 'food crisis' as an unintended result, or to blame it on higher oil prices or on the diversion of acreage to biofuel or the increased demand for soybeans in China is a cover up. Indeed, it is an exercise in

## US and EU government officials created the food price bubble

cynicism for the World Bank to wag its finger at governments promoting biofuel production, as it has recently done.[4] Biofuel production is perfectly in tune with its policy recommendations that have systematically prioritised the commercialisation of agriculture and profit maximisation at the expense of subsistence needs.

'Perfect storm', however, is not the only description that obscures the social agents responsible for the sudden food price increases. For the crisis is also often depicted as a price 'bubble' driven by the irrational speculative activity of short sighted investors who are bidding up wheat, corn and rice futures far beyond their 'true value' in a desperate desire to wring every last bit of profit before the bubble bursts and the price collapses. The implication being that these investors are as caught up in the process as the people who do not have the money to pay for the corn flour to make their daily tortillas! So, according to this logic, if a market bubble is responsible for the death by hunger and malnutrition of millions, then *nobody* is to blame.

The 'bubble' description of the price hikes is plausible because more and more aspects of capitalism in this neoliberal period are becoming 'financialised'. Thus, in the major mercantile exchanges for grains, investment funds and hedge funds have joined agribusiness and food processing firms as the major buyers and sellers. They are involved in these markets not in order to make a profit out of selling the commodity or through using it to produce other commodities, but in selling the right to sell the commodity at some fixed price in the future. This motivation creates the conditions for a bubble to develop.

The 'bubble' method of dramatically increasing the price of staple foods is quite different from the way it was done in the last historic increase of staple food prices in the early 1970s. As Harry Cleaver incisively describes it, the US government deliberately created a grain shortage by holding down acreage allotments for cultivation in 1970, 1971, and 1972 even after agreeing to the massive grain deal with the Soviet Union.[5] This led to a scarcity of grain on the interna-

## George Caffentzis

tional market and a dramatic increase in its price. This type of market manipulation was done during the Keynesian period when the state could justifiably claim to control the market; after all, this was the period of Nixon's wage-price controls. In the present neoliberal era this open planning would be ideological anathema, especially for the Bush administration, and so the grain price increases are accomplished through the market.

But we should not be deceived into thinking that 'bubbles' are unintended distortions that in time will be corrected and everything will go back to 'normal'. For bubbles are not irrational, though the investors involved in them might be; bubbles are constructed, inflated and 'popped' by financial institutions and the state (for a classic example see the role of Alan Greenspan's Federal Reserve in the 'dot.com' bubble of the late 1990s). They have their purposes and bankers, investment fund executives, and US and EU government officials are responsible for their creation and collapse, including that of the food price bubble of 2008.

## Food Power

If the 'food price crisis' is the result not of error and irrationality, but of planning and capitalist reason, what is its purpose? What objective/s is it supposed to achieve? More broadly, what is its significance from the viewpoint of the short and long-term trends in capitalist accumulation and class relations?

One sure answer to these questions is that the present food crisis is but the latest step in international capital's long march towards establishing its control over the main planetary sources of energy and value; food production being the key to the regulation of economic activities, wage levels, and labour power production in every part of the world. It is a long march that dates back to the colonial era, when not only were land and other natural resources expropriated, but export-oriented cash cropping and monocultures were implanted at the expense of the colonised people's subsistence needs. Since then, every new policy regarding agriculture and food production (at times involving 'cheap' and at other times 'expensive' food) has been shaped by the principle that *those who control food production also control the political economy of the planet*.

The drives to privatise communal lands and to industrialise agriculture through its rooting in the chemical industry and, later, biotechnology, from the Green Revolution to the invention of Genetically Modified (GM) seeds, were both geared to this purpose. So were the politics of agricultural subsidies and 'food aid' which the US (the Prime Mover in the commercialisation and corporatisation of global food production) has promoted since the 1950s. A turning point was the Food for Peace Act of 1954 that politicised food aid and began to make formerly colonised countries dependent on grain from the US. Much of this was achieved through

## Descrambling the 'Food Crisis'

the policy of 'cheap' food that, in time, undermined the livelihood of subsistence farmers in Africa, Asia and Latin America by making it impossible for them to compete with the heavily subsidised and industrialised agriculture of the US.

It is important here to stress that the use of food as a means to create dependence has been at the heart of the rise of agribusiness and the displacement of millions of small farmers also in the US. And it is a policy that has continued to the present as proven by the US government's staunch refusal to accept the request of many NGOs that they buy food with the money it 'donates' on the local markets from the farmers of the 'poor countries', rather than importing it from the US. This refusal has been so scandalous and so openly detrimental to the countries supposedly 'aided' that even a quasi-governmental NGO like CARE International, which has a deep allegiance to Washington, having worked for decades with the Pentagon in times of war, in 2007,

> rejected $46 million of food aid from the United States government because the aid was destroying the livelihoods of the very people, the farmers, it was meant to assist.[6]

## Price Crisis = Class War

Placing today's 'food crisis' against this historical background reminds us that capitalism produces scarcity rather than wealth at least for the majority of the world population. But this explanation also raises new questions. If capitalist development has for decades been structured by the need to control food supplies and use food as a means of profit making and as a weapon in the class struggle, why is it still necessary, in 2008, to resort to measures like dramatic food price hikes that are guaranteed to trigger revolts in many parts of the world? Why do the World Bankers and other capitalist planners want to run this risk? Why were the policies of structural adjustment not sufficient? And, is the food crisis part of the crisis of neoliberalism?

These questions are not easy to answer, but some hypotheses are in order. For the dramatic and murderous food price increases are intended to force a decisive shift in the constellation of class forces throughout the world in favour of capital.

First, *the food crisis should be read as a reply to the widespread refusal of land commodification (which is especially strong in Africa) and the struggle that communities are making across Latin America to reverse the privatisation of land and natural resources and recreate 'new commons'*. This has taken the form of open land re-appropriation movements, like the Zapatistas, the Landless Workers' Movement in Brazil, the Landless People's Movement in South Africa, as well as many less formally organised efforts throughout the world.[7] In many parts of Africa, rural migrants to the towns, especially women, have been cultivating plots of public land, a move which enables them to gain some independence, increase the family consumption and budget, and raise some money of their own through the small scale sale or barter of the surplus product.

Both World Bank officials and the CEOs of the major food trading corporations undoubtedly plan that the food price hikes will put an end to this resistance, as the rising food prices will produce a new 'land rush', leading to further land expropriation, a further commercialisation of agriculture, and renewed attacks on subsistence farming in favour of the farming of crops for export.

Second, through the hike in food and fuel prices, capital is trying to introduce a set of reforms in the social reproduction process that have long been on its neoliberal agenda, but so far have been successfully resisted by workers in Europe and the US. Despite the erosion of benefits due to increasing inflation, neither Social Security in the US nor pensions in many countries of the EU have been significantly reduced, despite repeated efforts to achieve this end. In this context, the food price hike is the equivalent of a cut in the real wage and a transfer of even more value to the agricultural corporations. It disposes of those 'income rigidities', to paraphrase a Keynesian concept, that so far have prevented the far-reaching 'welfare reform' capital has aspired to for many years. As with inflation historically, food price hikes attack working class communities at their weakest, as shoppers and consumers, rather than as members of workers' organisations.

Third, the food price hike also serves to defeat the resistance of many governments, from Europe to Africa and Latin America, to the introduction of GM

# those who control food production also control the planet's economy

products. The rejection of GM food has surprisingly come from every social rank in these regions, to the great dismay of US agribusiness. Significantly, the argument that hiking food prices would facilitate the African governments' acceptance of GM was already well articulated in the World Bank's World Development Report of 2008, that was published in October 2007 (before the most recent price hikes). Indeed, the report would appear quite prophetic of the developments to come, if we did not realise that to a great extent these developments prevailed thanks to the efforts of the World Bank and its partners in the UN system, the IMF and the FAO, 'prediction' often being a euphemism for a planning target in World Bank discourse.[8]

# The Future of Food Production

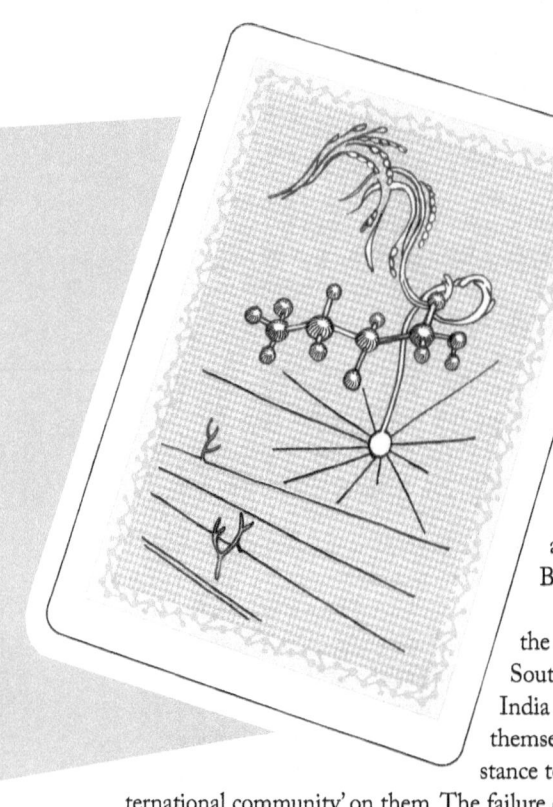

It is difficult to anticipate whether the food crisis will achieve these three objectives or will instead stimulate a worldwide uprising of movements for the re-appropriation of land and de-commodification of agriculture. For the best laid plans of both agribusiness executives and the World Bankers can go astray.

One determining factor here will be the behaviour of governments in the global South (especially the largest ones, China, India and Brazil), many of which have shown themselves ready to take a more combative stance towards the pressure exercised by the 'international community' on them. The failure of the World Trade Organization's 'Doha Round' is a good sign in this direction. But resistance is not confined to giant nations like China, India and Brazil. An exemplary precedent is the case of Malawi, one of Africa's smallest countries. Traditionally self-sufficient, and even an exporter of food to South Africa, in the 1980s this country was led by the World Bank to end its subsidies to farmers and later, in 2000, the IMF insisted that it put its food reserves on the market. But after years of near famine conditions, it has recently reversed these policies, despite pressure to the contrary by the IMF and World Bank. This has been a wise move, indeed, one that other countries will likely emulate, since for the first time in years Malawi can boast again of a surplus food production.

States, however, are problematic allies in the struggle against starvation-creating price hikes, since they are almost always also interested in developing their capitalist agricultural sector, *but on their own terms*. What will matter most in deciding the outcome of the food crisis will be the ability of movements, across the divides imposed on us, to coordinate strategies of resistance against the planners of hunger and starvation. The struggle against famine and land expropriation cannot be fought only in Africa or the mountains of Chiapas. It must be carried on in the supermarkets and streets of the US and Europe, as were the 'clean clothes' campaigns of the 1990s that transformed shopping into a political act and united textile sweatshop workers in the Global South with workers in the North. The demand for food for all, that will strengthen not poison the consumers and producers,

George Caffentzis

## The 'food for all' demand founds a material politics against crisis

forms the basis of a material politics that can upset the purposes of those who have planned the food price crisis.

But let us make no mistake on this point: the stakes in this struggle are high. If the food price hike strategy works and food production is fully commercialised, if the struggle for the re-communalisation of land is defeated, with food and land going to those who have more cash in hand, millions will die, the peasantry as a historic class will disappear, people across the planet will have access to land only by labouring as peons at the service of agricultural corporations, and we will have missed our last chance of having some control over the quality of the food we eat, which is already largely a byproduct of the petro-chemical industry. Most important, if the food price hikes achieve their purpose, the possibility of constructing a world where, paraphrasing José Bové, 'Life is not for sale', will be extinguished for decades.

### Footnotes

**1** See: Maria Mies, *Women, Food and Global Trade: An Ecofeminist Analysis of the World Food Summit, Rome, 13-17 November 1996*, Bielefeld: Institute für Theorie und Praxis der Subsistenz e.V., and Karen Lehman & Al Krebs, 'Control of the World's Food Supply', in Jerry Mander and Edward Goldsmith (eds.) *The Case Against the Global Economy and For a Turn Toward the Local*, San Francisco: Sierra Club books, 1996.

**2** Mariarosa Dalla Costa, 'Food Sovereignty, Peasants and Women', *The Commoner*, 2008, http://www.thecommoner.org

**3** Walden Bello, 'Manufacturing a Food Crisis', *The Nation*, 2 June, 2008.

**4** World Bank, 'World Bank Development Report for 2008: Agriculture for Development'. Washington, D.C.: World Bank, 2007.

**5** Harry Cleaver, 'Food, Famine and the International Crisis', *Zerowork* II, 1977, pp.7-70.

**6** Kwesi W. Obeng, 'Soaring food prices send shockwaves and protest across Africa', *African Agenda*. Vol. 11 No.1, pp.5-9.

**7** Sam Moyo and Paris Yeros (eds.), *Reclaiming the Land: The Resurgence of Rural Movements in Africa, Asia and Latin America*. London: Zed Books, 2005.

**8** World Bank, op. cit.

> George Caffentzis <gcaffentz@aol.com> is a member of the Midnight Notes Collective. Together with the collective, he has co-edited two books, *Midnight Oil: Work Energy War 1973-1992* and *Auroras of the Zapatistas: Local and Global Struggles in the Fourth World War*. Both were published by Autonomedia Press

# FROM SUBPRIME TO SLUMP?

This year the world has seen the power of money to socialise the costs of capitalist crisis, but are prices going to go on rising to Weimar-like levels? <u>Jon Amsden</u> explores the origins of the crisis and discerns something worse than inflation on the horizon

In May of this year, Brian Marks made a valiant attempt to tie together inflation, the current crisis in financial markets, and struggles of the world working class. Marks wrote:

> The food and energy crises are key ways capital is trying to displace the costs of devaluation onto the working class. (Foreclosures, the manipulation of interest rates, and the outright bailout of banks with public money are other important measures). The transfer of workers' wealth through energy and food costs to the energy sector is then conveyed in a concentrated form to save (by buying up) the banks in crisis. That is where primitive accumulation meets fictitious capital.
> – Brian Marks[1]

Marks argued that inflation is a special form of 'looting' whereby the capitalist class attempts to appropriate 'the wealth of the workers', for the purpose of 'propping up fictitious capital'. In a discussion on the Meltdown mailing list Ben Seymour queried Marks' logic on this point, noting that since inflation essentially devalues the workers' portion, at least in monetary terms, such would not be a particularly effective form of 'looting'. For Seymour, the very thing that is supposed by Marks to constitute an accelerated 'looting' of the working class, 'an escalation of the ongoing compulsion of work' which presumably increases the rate of surplus value extraction, is at the same time undermining or cancelling out the value extracted. Thus, crisis can increase the economic pressure on the working class, but the actual rate of exploitation is offset by the devaluation of the currency which measures the product and price of their labour power. We will visit the 'increased economic pressure' that inflation places on the working class a bit later on. In one respect, however, inflation can, in fact, be favorable to that part of the working class who may be net debtors. For example, Joe Sixpack has an outstanding credit card balance of $9,000 US. As the real value of this figure is diminished by inflation, Joe will have to contribute less value to pay it back than he received (on credit) in the first place.

Both Marks and Seymour were writing in the attempt to make the current crisis understandable within the familiar paradigms of Marxist theory. There are potential problems here because, as I believe, Marxists do not exactly speak with one voice as to the ultimate causes of capitalist crisis. In his Penguin edition of Marx's *Capital* Vols. I-III, for example, Ernest Mandel tells us that, 'Marx did not have a theory of crisis.' Currently, Marxists tend to argue vigorously over Marx's suggestion that the final cause of the cycle of boom and bust that has typified capitalist economies since the beginning is a tendency for the rate of profit to fall.

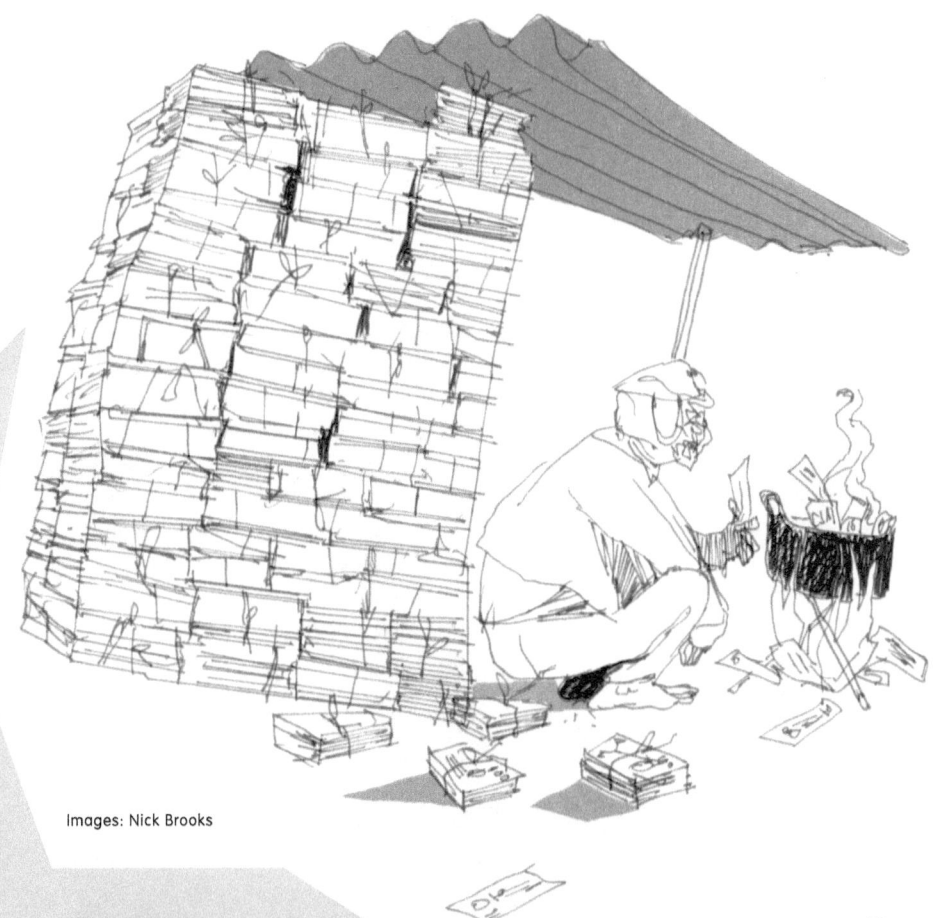

Images: Nick Brooks

**From Subprime to Slump?**

This is another topic to which we shall return after setting the stage with the tragedy (and the folklore) of the current crisis. The first bit of stage setting has to do with the reality of the rate of change in prices (inflation or deflation).

When one attempts to make sense of the supposed causes of the current crisis, one finds that the rate of change in prices noted by Marks (Brian not

## economic gurus in finance, government, and media have got the whole thing back to front

Karl) makes up an important element in analysis of the ongoing collapse of financial markets and the underlying slowdown of the real economy. The focus on inflation is found in equal measure whether we read the financial press, scan the internet blog inferno, or simply talk among ourselves about what is going on in financial markets right now. The big question about the present chaos, both in financial markets and in the real economy, is whether or not it is the credit crisis which caused the slowdown in economic activity or conversely if it is the economic slowdown that has given rise to the credit crunch, to bank and corporate failures, to the disappearance of consumer confidence, to ever-increasing levels of unemployment and all the rest of it. The view that dominates the bourgeois media at the moment is that it is the collapse of the housing bubble and the attendant failure of financial institutions that has caused the 'recession'. In my view, the economic gurus of high finance, the government, and the media have got the whole thing back to front.

Since the main economic remedy initiated and celebrated both in academic and Wall Street circles is to pour more money on the problem, the role of inflation in the current mess becomes a central one. What most of us want to know now is whether or not the inflation that has marked the early moments of this crisis will be maintained throughout the coming collapse of capital markets. Why do we care? Let's pause a moment and look at what happens when inflation is rising. First, your life changes in a dramatic way. Americans who dearly love their oversized gas guzzling automobiles are learning to stay home more. Perhaps more important is that those who have any savings or hold any cash at all are watching it evaporate day by day. It wasn't that easy to earn it and now it is gracefully disappearing as the general price level keeps rising. Those lower down on the earning scale whether in the US, the UK, or elsewhere are learning to change their diets and to give up beer (even on weekends). What this all adds up to is a growing sense of

Jon Amsden

insecurity and anger and (eventually) the search for a social scapegoat. German Nazism, for example, traced part of its heritage to the runaway inflation that plagued the Weimar Republic. On a less dramatic level, the way that inflation will affect your life is that you will discover a growing eagerness to spend the money you have on what you need or want before prices rise. It is possible that chronic inflation is one of the most serious economic phenomena that exists. But it is unclear whether inflation will continue to be the dominant form of the ongoing economic crisis.

Consequently, what people want to know at this point is whether or not the present positive rate of change in prices (inflation) will continue to accompany all of the ills of 'recession' that we have come to know and fear. These include: losing our jobs, losing our savings (Indy Mac, Northern Rock), watching the value of houses and apartments plummet, the breakdown of international trade (the 'stalled' Doha Round) and, in general, the social and economic turmoil that lies ahead. The answer proposed here is that, no, inflation as we know it will not continue to dominate the reality of and discussions about the ongoing economic crisis. On the contrary, the current inflation (sometimes called 'stagflation') will become deflation (the rate of change in prices goes negative) and, as past history shows, the deflationary crisis could be infinitely worse than what we are going through at the moment.

As a way of getting into the current discussion of these matters, let's take a look at the financial and international press on a particularly confusing and anxiety provoking weekend, 16 August, 2008. The eminently respectable *Financial Times* (FT) tells us in its front page headline that there is a 'Surge for the Dollar as Global Fears Rise':

> Against sterling, the US currency notched up its 11th consecutive day of gains – its longest uninterrupted rise in more than thirty-five years – as markets became increasingly convinced that the US was best placed to weather the global downturn.

The dollar's rise, according to the FT, was 'triggered' by a sudden collapse of commodity prices. What the FT is trying to say, in simple language, is that when the commodity price bubble suddenly deflated, the dollar became *ipso facto* more valuable. What the FT then asserts, in a more or less imperious assumption, is that the sudden 'surge' in the value of the dollar reflects a worldwide (and to my mind extremely unlikely) growth of confidence in the US economy.

'Confidence?' The US domestic and trade deficits are in the hundreds of billions for any foreseeable future, major banking and financial institutions are being handed billion dollar crutches by the Federal Reserve, the real economy is grinding to

**From Subprime to Slump?**

a halt, all the Buicks are now made in China, and the FT proclaims 'confidence?' Please, somebody get me a glass of water, I have to sit down for a moment or two! Either that or hand me a copy of the weekend *International Herald Tribune* so that I can regain a sense of balance!

The 'Weekend Business' section of the *International Herald Tribune* (IHT) for 16 August features one of the great economic naysayers on the 'other side of the pond' who, under no less than four reiterated photographs of himself looking 'worried', is headlined as follows: 'The Seer Who Saw the Storm Coming: Professor Warned of Financial Crises One Year Before They Struck.' The IHT, in the person of junior professor, Stephen Mihm, then has lunch with the famous/infamous Nouriel Roubini of New York University's Stern School of Business who is, well, 'very worried'.

Roubini is the sole proprietor of a website entitled 'RGE Monitor' [http://www.rgemonitor.com/blog/roubini/], which regularly provides pessimistic arguments on the US economy to the following constituencies: bear market enthusiasts, students of the international economy and those members of the US Left (including me) who would only be too happy to see the whole overpaid, hyper-

# the economy grinds to a halt, all the Buicks are made in China, and the US is 'confident'?

inflated, double-talk infected, economically exploitative, and (more or less) criminally inclined financial sector of the US (and global) economy fall flat on its face, never to rise again. Mihm says, in reverential tones, that, from the screens of the RGE Monitor, Roubini called the ongoing financial crisis of 2008 exactly one year before. Unfortunately, nobody listened.

Just between us, the man who was christened 'Dr. Doom' by the more light-hearted *New York Times* was not the *only* prophet of doom. There were several others who explained their position a bit more clearly than Dr. Doom, but to make sense of it, a brief reminder is necessary. The financial community, as most people know, is divided between 'bulls' (optimists) and 'bears' (pessimists). However, what most people don't know is that the bulls and the bears speak for different financial constituencies that divide up rather neatly along the lines of how much money the folks concerned have to invest. The bulls commonly speak for (and to) investors in equities (i.e. shares) including small investors, those with their life savings marooned in pension plans, and (of course) widows and orphans.

Jon Amsden

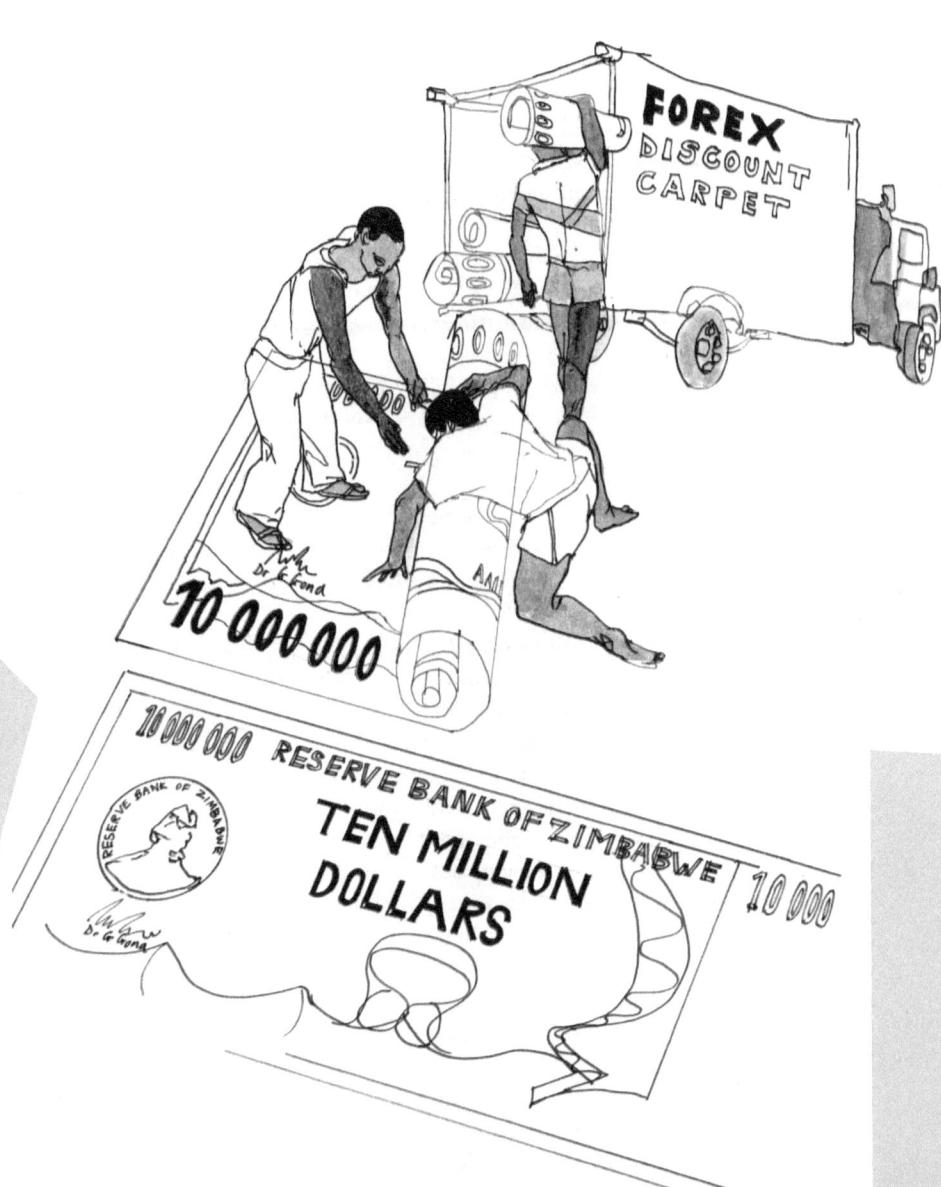

## From Subprime to Slump?

The bears, on the other hand, commonly speak to the investors who have serious money to place and who tend to prefer bonds of all kinds, but especially government bonds. What divides the 'bulls' and the 'bears' (sometimes called 'the bond gods') is guess what... fear of inflation. Fixed income securities become less valuable in an inflationary scenario. The 'bond gods' tend to equate inflationary pressures with a vibrant market for equities, so there is often more than a little *Schadenfreude* on display when equities take a hit and bonds are doing better. The professional bears tend to have more complex and powerful arguments about economic decline than does the NYU professor mentioned earlier.

Roubini has been concentrating on one single theme of the coming disaster, namely, the huge current account deficits typically run by the United States. The current account deficit measures what America *does not pay for* in exports in terms of what it cheerfully continues to consume in imports. For the last few years, the US current accounts deficit ran at about 1 billion dollars per day, but recently this has risen to much higher figures. By concentrating on this grim economic variable, Mihm asserts, Prof. Roubini could foresee:

> ... a bleak series of events: homeowners defaulting on mortgages, trillions of dollars of mortgage backed securities unraveling worldwide, and the global financial systems shuddering to a halt.

Quite how Roubini can get from a negative trade deficit to this varied and complex list of 'worries' is not explained. It is, in fact, unlikely that anyone could deduce or predict all of the above troubles from the simple fact that the US is wildly overdrawn on its credit card. Never mind!

Roubini, because he is 'worried', and because he wants to save the market capitalist economy from itself, has applauded the various actions of the Federal Reserve intended to stave off economic ruin. This is where the question of inflation, or 'stagflation' (prices keep going up as the economy winds down) comes in. Why did Roubini applaud loudly when the US central bank ('the Fed') cut interest rates to save the US economy? The constant assumption here is that cheaper money will stimulate the capitalist production machine to go back to work. Roubini also endorsed the Fed's intervention to prevent the total collapse of Bear Stearns, investment banker, by (essentially) nationalising it with the help of J.P. Morgan. Now J.P. Morgan, the biggest, oldest, and richest investment bank in the US, is writing off billions of dollars – but not to worry! Both actions applauded by our 'worried' professor put more 'money' into the system at a time when the economy was slowing down. More money alongside fewer goods being produced usually spells inflation.

Jon Amsden

This is, of course, precisely what happened and will probably continue to happen for a while. What *did not* happen, however, was that injections of cheap money restarted the economy. The reason is that, despite everything you have read, the sudden shortage of bank credit was not the *cause* of economic slowdown but, rather, its *effect*. It was the fact that people were losing their jobs and couldn't pay their mortgage payments due to the economic slowdown that caused the housing crisis and not the other way round. Thus, the mortgage crisis happened (all over the world) because many mortgage holders could no longer make their mortgage payments. Why not? Because the working class were either losing their jobs, going on short time, or taking lower paid jobs just to get by. The question we now need to look at is why do economies slow down (or die altogether) and what happens to the rate of change in prices (inflation or deflation) when they do?

The classic Marxist position is that both an expansion and a contraction in capitalist economies are needed for the accumulation of capital. In the expansion phase, a new technology is adopted and the first comers make a lot of money, wages are bid up, and the prices of goods fall. In the US, a classic expansion of this nature took place, for example, after the Second World War. War production

# the sudden shortage of credit was not the cause of economic slowdown but its effect

had decreed the introduction of new technology, workers had organised unions to demand higher wages and (briefly) to ascend to a middle class lifestyle. The jobs they held, by the way, had medical coverage, pensions plans, and secure employment until pensions kicked in. That world is now gone forever.

Going back a few years, again for the American case, the Great Depression (1929-1939) was a classic contraction. Banks failed by the thousands, industries closed, something like 25 percent of the workforce was without jobs, and (this is the interesting part) prices of everything fell through the floor. This was the classic deflationary crisis, i.e the sort of crisis that had characterised capitalist development from the beginning. In Marxist theory, the importance of deflationary crises is that fictitious capital (largely overbid stock prices) is destroyed, larger and more efficient capitalists gobble up the little ones thus laying the ground for the introduction of new technology, and (most important) the working class is disciplined for the next round of production.

## From Subprime to Slump?

Both the expansion and the contraction phase are necessary for more and more capital to be accumulated, so Marxists call this process the 'cycle of capital accumulation'. In the classic Marxian doctrine, the downturns are caused in two ways. The small ones are caused by marked instability between the industries that produce consumer goods and those that produce capital goods (*Capital* Vol.II). The big ones are caused by a long run tendency of the rate of profit of capitalist enterprise to fall.

The rate of profit falls as technological innovations spread throughout industry. The first comers have already become rich, but when the whole sector adopts a new technology (say, steam power to replace water power), profits are eventually competed away. This is at the macro level. At the level of the firm, Marx argued (*Capital* Vol.III), the change in the labour to capital ratio in favour of capital diminishes the amount of living labour time that the capitalist accumulates. Then, since living labour is the source of all economic value, the rate of profit tends to fall.

The picture presented by Marx may seem counter-intuitive, but if you look at the world today you will see the results of this process that are, once again, manifesting themselves. Now capital has fled the highly industrialised regions of Europe and North America to be applied in China, Vietnam, and other relatively backward economies. Why? Because, of course, this is where the capitalist gets to accumulate most prodigiously the source of all surplus value which is human labour time.

How, then, did the present US and world crisis begin this time around, and will it continue in an inflationary or deflationary context? Let's close with this. As the rate of profit declines in the older sectors of capitalist enterprise, the so-called 'financialisation' of the economy begins. The older metropolitan regions no longer export goods to the rest of the world. They now export capital. Lenin first noticed this process as the European countries began to export capital to other regions towards the end of the 19th century. Great Britain, for example, built power plants and street railway systems all over the world in this period. A similar process is taking place today, but the result now seems to be that the productive sectors of the mature economies are running down just as the financial sectors are becoming wealthier than ever. The result is 'funny money games' in financial markets, including the creation of useless economic derivatives, the 'lending' of capital to developing nations and the hectoring of less developed nations by the IMF to make them pay their 'debts'.

This is pretty funny. The US is the largest debtor nation on the planet as we speak, but its financial geniuses (until lately) have been 'lending' through the World Bank and the International Monetary Fund. How do they do this? Well, it can only happen in a world in which the international reserve currency (currently the US dollar) is created by simply saying that it exists. This bit of magic has allowed the US to 'live beyond its means' for some time now. It has also put the Chinese and the Indians in the grotesque position of financing the US wars for petroleum along with an out-of-control lifestyle obtained on credit. As a system, what could be called 'dollar imperialism' is entering the disaster stage, both with respect to the costs and risks of war and with respect to the collapse of the real economy in the UK, the EU, and the good old USA.

Given these realities, will the present inflationary moment continue until we reach the point of total collapse? I happen to think not, though others whose views I respect point to the dangers of runaway inflation which are far more socially disruptive (think Weimar and the growth of Nazism). In my view, what will happen next is that, when the US economy completely collapses in both the financial and real sectors, the total collapse of the banks will follow with widespread corporate failure and the increase of unemployment to unbelievable levels. Under these circumstances demand will fail (as in the last great deflationary crisis of 1929-39) and prices will fall through the floor.

There is another scenario, however, namely that of runaway inflation. Explaining this would require an extensive discussion of the role of the dollar as the international reserve currency. Briefly, the collapse of the dollar internationally could lead to runaway inflation in the whole global economy, with awful social and political consequences. One would have to take a more careful look at one of the central concerns of Professor Roubini, namely, his obsession with the American tendency to run a current account deficit worth $2 billion a day(!), but this is a topic better left for another time.

### Footnotes

1 Brian Marks, 'Living in a Whirlwind, or the Food/Energy/Work Crisis',
http://www.journalofaestheticsandprotest.org/contents.html

For some criticism and discussion of this text, see Ben Seymour and Jon Amsden:
http://www.metamute.org/en/living_in_a_whirlwind_or_the_food_energy_work_crisis_and_some_criticisms

Jon Amsden <thewriterscoach@gmail.com> PhD. LSE grad. Old Left. Soixante-huitard

# MR SMITH GOES TO BEIJING

Sociologist Giovanni Arrighi invokes the political economy of Adam Smith to claim that China's 'labour intensive' mode of production is the future of capitalism. It's also the past, argues Daniel Berchenko

In 1977, the *New Left Review* devoted more than half of its July-August issue to an essay by historian Robert Brenner entitled 'The Origins of Capitalist Development: A Critique of Neo-Smithian Marxism'.[1] The target of Brenner's critique is Immanuel Wallerstein's *The Modern World System* – a book that is generally credited with (or in this case, accused of) laying the foundations for the interpretive framework known as 'World Systems Theory'.[2] World Systems Theory is an unlikely synthesis of Marxism, Annales School historiography, and Dependency Theory that was sweeping sociology departments during the late 1970s. The novelty of the approach consists in viewing the entire history of capitalist development over the course of the last 500 years as a unified process whose subject is not any one class or nation, but the global market itself (the eponymous 'world system'). Wallerstein's thesis is that the emergence of capitalism and the continual revolution of the means of production are driven by the dynamics of international trade. 'Capitalism and a world economy (that is, a single division of labor, but multiple politics and cultures) are obverse sides of the same coin.'[3]

In his critique, Brenner compares Wallerstein's model to the political economy of Adam Smith in *The Wealth of Nations*. Like Smith, Wallerstein argues that the expansion of trade creates conditions under which goods are increasingly produced for exchange, rather than for immediate consumption. Like Smith, Wallerstein believes that this situation leads directly to the development of the forces of production and the accumulation of capital. For Wallerstein, production for exchange constitutes the *differentia specifica* of capitalism as mode of production:

> The essential feature of a capitalist world economy ... is production for sale in a market in which the object is to realize the maximum profit. In such a system, production is constantly expanded as long as further production is profitable, and men constantly innovate new ways of producing things to expand their profit margin.[4]

In the models of both Smith and Wallerstein, the profit motive compels individual capitalists to specialise in particular branches of industry and to rationalise their

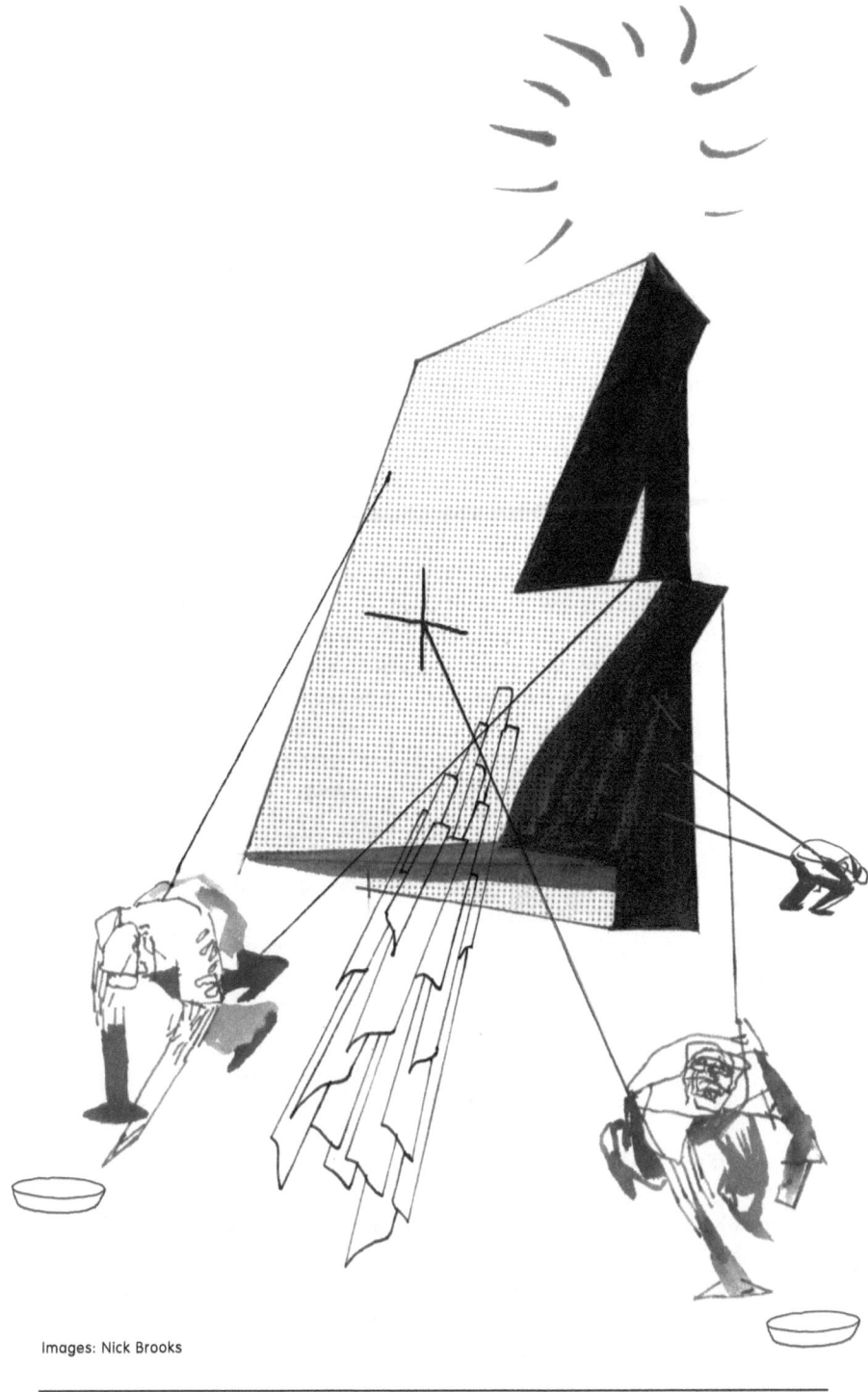

Images: Nick Brooks

production processes. The progressive accumulation of capital in these industries raises barriers to entry, diverting subsequent investment into new channels. The result is a social division of labour whose depth is determined by the extent of the market. Smith and Wallerstein differ only in their estimations of the salutary effects of this process. (Whereas Smith notoriously viewed the expansion of trade as the path to international peace and prosperity, Wallerstein argues that the structure of the modern world system benefits the core at the expense of the periphery.)

Brenner takes issue with Wallerstein's suggestion that production for exchange leads directly to the emergence of capitalism and capitalist development. He notes that exchange-based production existed in many pre-capitalist societies before taking root in early modern Europe.[5] Because pre-capitalist societies are fundamentally agrarian, both exploiters and direct producers have access to their own means of subsistence. '...[T]heir survival and reproduction is not dependent on the sale of their products on the market,' writes Brenner, 'consequently they do not have to compete in terms of productive powers.'[6] Under these conditions, 'the market exerts no pressure toward the *continual* revolution of the means of production.'[7] As a result, there is a bias in pre-capitalist societies toward the realisation of 'absolute', as opposed to 'relative', surplus value. In other words, because labour is compulsory for serfs and slaves, lords and masters tend to extract additional surplus labour by lengthening the working day or extending the corvée, rather than by introducing technological innovations. There is little reason to invest profits in the development of productive forces. 'Rather than being accumulated, economic surplus is here systematically diverted from reproduction to unreproductive labour.'[8] According to Brenner, '[t]he increase of relative surplus labour cannot become a *systematic* feature of such modes of production.'

Brenner, following Marx, argues that capitalism emerges only when labourers are both free to sell their labour power on the market as a commodity, and compelled to do so in order to survive:

> As Marx puts it, 'the domination of exchange value itself, and of exchange value producing production, presupposes alien labor capacity itself as an exchange value – i.e. the separation of living labour capacity from its objective conditions ...'.[9]

This point is central to Brenner's critique. For Smith and Wallerstein, capitalist class relations are the *result* of the expansion of trade and the deepening of the division of labour. According to their model, capitalists are driven by the profit motive to innovate and develop the forces of production. Beyond a certain point, this development requires a reorganisation of labour within the individual productive unit. Smith's pin

Daniel Berchenko

factory must be staffed by wage labourers from the semi-educated urban working class, rather than by uneducated serfs. Thus, capitalist relations of production gradually supplant feudal relations with the development of trade and market-based production; wage labour autonomously replaces the corvee. Against this view, Brenner argues that capitalist class relations are a necessary *precondition* for the accumulation of capital.

> ... 'production for profit via exchange' will have the systematic effect of accumulation and the development of the productive forces only when it expresses certain specific social relations of production, namely a system of free wage labour, where labour power is a commodity. Only where labour has been separated from possession of the means of production, and where labourers have been emancipated from any direct relation of domination (such as slavery or serfdom), are both capital and labour power 'free' to make *possible* their combination at the highest level of technology... Only under such a system, where both capital and labour power are thus commodities – and which was therefore called by Marx 'generalized commodity production' – is there the necessity of producing at the 'socially necessary' labour time in order to survive, and to surpass this level of productivity to ensure continued survival.[10]

Brenner therefore accuses Wallerstein and Smith of employing a kind of circular reasoning. The validity of their model is premised on the prior emergence of capitalism and capitalist social relations – the very phenomena that it seeks to explain.

Although more than three decades have passed since the publication of Brenner's essay, the Wallerstein-Brenner debate continues – somewhat in the manner

# 'labor-intensive production' cannibalises existing resources until it literally devours itself

of a summer movie franchise, whose first installment has been forgotten by most of its audience, and many of its current participants. This may explain why few expressed surprise recently when Giovanni Arrighi, Wallerstein's most prominent acolyte, published a book in which he endorses Brenner's position in the NLR essay and self-consciously claims the Neo-Smithian title for his cause. In *Adam Smith in Beijing: Lineages of the Twenty-First Century,* Arrighi takes up Brenner's distinction between 'Smithian' and Marxist models of development in order to analyse the so-called 'economic renaissance of East Asia.'[11] Arrighi concedes that exchange-based production does not lead directly to the emergence of capitalism. However, rather than concluding with Brenner that the Smith/Wallerstein model is a theoretical fiction, he argues that it constitutes a viable alternative to capitalist development.

Arrighi claims that a form of 'Smithian' market-based development occurred in China between the 16th and 18th centuries. During this period of relative peace and stability, China experienced massive population growth, paralleled by an expansion of its domestic market. However, these conditions did not result in any corresponding development of the means of production. Arrighi quotes a passage from a study by Mark Elvin, describing some of the fetters to technological development in China at this time:

> [W]ith cheapening labor but increasingly expensive resources and capital, with farming and transport technologies so good that no simple improvements could be made, rational strategy for peasant and merchant alike tended in the direction not so much of labor saving machinery as of economizing on resources and fixed capital.[12]

According to Arrighi, these tendencies led to the development of 'labor-absorbing institutions and labor-intensive technologies' that allowed demographic growth to continue in China, in spite of natural resource constraints.[13] Arrighi follows World Systems fellow-traveller Kaoru Sugihara in referring to this process as an 'Industr*ious* Revolution'.[14] For both Arrighi and Sugihara, the Industrious

# What *are* the causes of the 'Chinese economic miracle', if not workers' industriousness?

Revolution is 'a market-based development that had no inherent tendency to generate the capital- and energy-intensive development path opened up by Britain and carried to its ultimate destination by the United States.'[15] Unlike the Western 'industrial' path, the East Asian 'industrious' path is characterised by a disposition to 'mobilize human rather than non-human resources.'[16]

Not surprisingly, critics like Robert Brenner and Philip C.C. Huang have challenged this interpretation of East Asian development during the early modern period. For Huang, the tendencies that Arrighi and Sugihara describe are fundamentally involutionary. Because the Chinese demographic boom of the 17th and 18th centuries was sustained by the increasing exploitation of the peasantry without any investment in means of production, it amounts to 'growth without development'.[17] (One example of industriousness, frequently cited by Arrighi, is

> the absorption of sideline non-agricultural work performed by women, children, and the elderly, reducing the operating costs of household production units, giving them a competitive edge over large capitalist units using hired labor.[18])

Brenner and his colleague Christopher Isett have argued that in the absence of any development of the forces of production, population growth is inherently limited and becomes increasingly precarious:

> In the Yangtzi delta, the main economic agents possessed direct non-market access to the means of their reproduction. They were therefore shielded from the requirement to allocate their resources in the most productive manner in response to competition. As a result, they were enabled to allocate their resources in ways that, while individually sensible, ran counter to the aggregate requirements of economic development, with the consequence that the region experienced a Malthusian pattern of economic evolution that ultimately issued, in the eighteenth and nineteenth centuries, in demographic-cum-ecological crisis.[19]

According to this view, 'labor-intensive production', as Arrighi euphemistically describes it, does not provide the basis for the extended reproduction of society. It cannibalises existing resources until it literally devours itself – leading to a Malthusian catastrophe.[20]

Arrighi rejects these criticisms as inherently chauvinist. He argues that Brenner and his allies unfairly measure the East Asian path against the standard of Western capitalist development.

> If we identify 'evolution' and 'development' as the displacement of labor intensive household production by capital-intensive production in units employing wage-labor, as Huang and Brenner do, then [the Industrious Revolution] should indeed be characterised as 'involutionary.' But if we leave open the possibility that labor-intensive production may play a lasting role in the promotion of economic development ... then such a characterization is unwarranted.[21]

As the passage above suggests, this is not merely a historical dispute for Arrighi. One of the central claims of his book is that China owes its recent fortunes (or misfortunes, depending on your view) to its industrious past.

> The East Asian economic resurgence has ... been due not to a convergence towards the Western capital-intensive, energy-consuming path but to a fusion between that path and the East Asian labor-intensive, energy-saving path.[22]

Arrighi cites, as an example, the Wanfeng automotive factory near Shanghai, where 'there is not a single robot in sight'. 'As in many other Chinese factories,' writes Arrighi,

> the assembly lines are occupied by scores of young men, newly arrived from China's expanding technical schools, working with little more than large electric drills, wrenches and rubber mallets.[23]

(Thankfully, the factory has yet to employ children and the elderly). In this Smithian paradise,

> Engine and body panels that would, in a Western, Korean or Japanese factory, move from station to station on automatic conveyors are hauled by hand and hand truck. This is why Wanfeng can sell its handmade luxury Jeep Tributes in the Middle East for $8,000 to $10,000. The company isn't spending money on multi-million dollar machines to build cars; instead, it's using highly capable workers [whose] yearly pay ... is less than the monthly pay of new hires in Detroit.[24]

Yet Arrighi maintains that the chief competitive advantage of Chinese manufacturers is not low wages per se, but rather 'techniques that use inexpensive educated labour instead of expensive machines and managers.'[25]

> ...[S]tatistics showing US workers in capital-intensive factories to be several times more productive than their Chinese counterparts ignore the fact that the higher productivity of US workers is due to the replacement of many factory workers with complex flexible-automation and material-handling systems, which reduces labor costs but raises the cost of capital and support systems. By saving on capital and reintroducing a greater role for labor, Chinese factories reduce this process.[26]

Daniel Berchenko

The reader is left to wonder why Detroit squandered so much money on conveyor belts in the first place. Some would argue that Western manufacturers invest in plant equipment for the express purpose of minimising labour inputs, since reductions in wages tend to offset the initial costs of investment over time. But more importantly, each manufacturer must achieve the same general level of productivity as its competitors (all things being equal – as we will see). Otherwise it will be forced to choose between selling its products at higher prices or achieving less than the average rate of profit. Therefore, when one capitalist invests in new plant equipment that allows it to produce more goods with less labour, its competitors must eventually adopt this new technology or perish. Marx describes this logic in the first volume of *Capital*:

> The law of the determination of value by labour-time makes itself felt to the individual capitalist who applies the new method of production by compelling him to sell his goods under their social value; this same law, acting as a coercive law of competition, forces his competitors to adopt the new method. [...] *Capital therefore has an immanent drive, and a constant tendency, towards increasing the productivity of labour...* [Italics mine][27]

Arrighi implies that because of their industrious past, Chinese manufacturers are able to achieve productivity gains *without* actually investing in plant equipment. But it is obvious that if this were possible, each advance would be universally adopted by competing manufacturers since, by definition, it would involve no investment in machinery or infrastructure. Arrighi himself provides evidence that the educational

# The reader is left to wonder why Detroit squandered so much money on conveyor belts

level of the average worker in China is well below that of the average worker in the United States. Therefore, any competitive advantage that Chinese manufacturers possess in terms of productivity would have to be purely organisational in nature. And while there can be little doubt that advances in organisational techniques have played a leading role in the development of the forces of production throughout history (think of the impact of Taylorism, for instance), it is equally certain that they are quickly disseminated and easily adopted when they do not require any corresponding investment in means of production. In other words, if automobiles

## Mr Smith Goes to Bejing

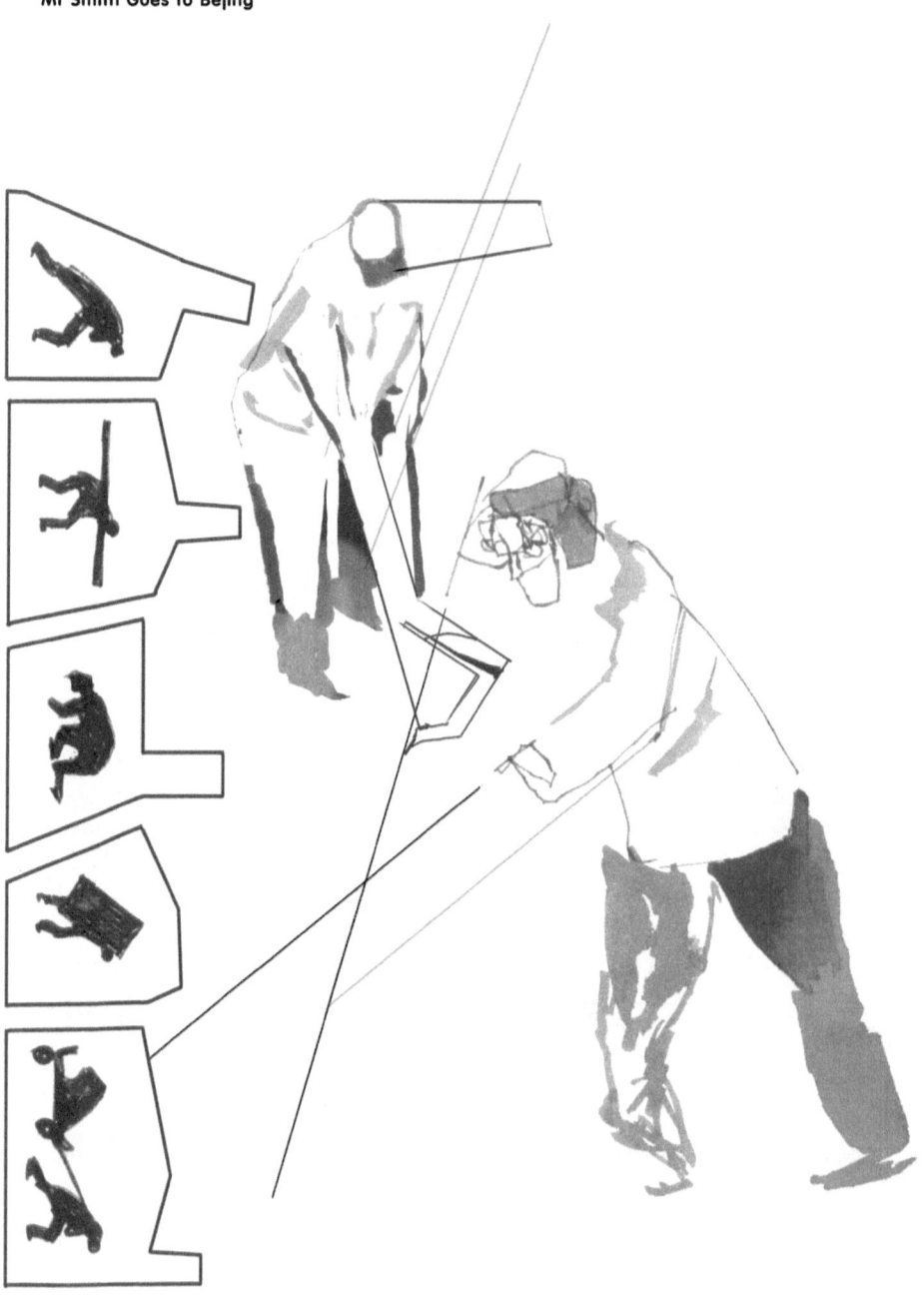

could be produced more efficiently with drills and rubber mallets than with automated machinery, workers across the globe would suddenly be possessed by the same spirit of industriousness as their Chinese counterparts. Arrighi's argument seems more indebted to the myth of John Henry than the political economy of Adam Smith.**28**

What *are* the causes of the 'Chinese economic miracle', if not the industriousness of Chinese workers? In order to answer this question, we must begin by inquiring into the nature of the miracle itself. Of course, every schoolboy knows that there are no miracles. But, at least in the case of the Chinese economy, this is one possibility that Arrighi fails to consider. After a few obligatory references to GDP figures and *Wall Street Journal* op-ed pieces, we are assured that

> China has increasingly replaced the United States as the main driving force of commercial and economic expansion in East Asia and beyond.**29**

Arrighi sets out from the assumption that China is undergoing some sort of 'economic renaissance' and proceeds in search of its causes. This leads to the absurd conclusion that a sweatshop in China is more efficient than a typical automated Western plant because of the 'industrious' heritage of Chinese workers (not to mention all of the money that those thrifty Chinese capitalists save by forgoing conveyor belts).

Is it possible that Chinese manufacturers are, in fact, less productive than their Western counterparts (even taking into consideration the cost of plant equipment)? If this were the case, how would we account for China's rising GDP and the explosive growth of Chinese exports in recent years? We noted above that manufacturers must achieve the same general level of productivity as their competitors – *all other things being equal*. However, Chinese manufacturers enjoy many unique advantages that allow them to prosper, even while employing primitive production techniques. Arrighi notes that in China, many of the social costs of reproducing labour power are borne by the village and the extended family. These informal support networks provide care for the sick and the elderly, as well as education for the young. They are also increasingly called upon to maintain roads, dispose of waste, administer civil law, and supply water and electricity – especially in the many slum-like villages that have been engulfed by China's rapidly expanding cities.**30** In addition, Chinese manufacturers are able to defer many of the social costs of the production process itself, resulting in environmental depletion and infrastructural decay. In the West, capitalists bear these costs to a much greater degree – either directly, in the form of wages and benefits, or indirectly, in the form of taxes and operating expenses.

By exploiting these advantages in order to cut prices and drive export growth, China generates enormous amounts of credit. Much of this credit returns to the US in order to subsidise its massive balance of payments deficit (creating equity and real estate bubbles in the process). However, China also uses its fortunes to subsidise many of its industries and to maintain cheap credit domestically. Cheap credit allows redundant and unprofitable businesses to survive by pyramiding bad debt. (Since much of this debt will never be repaid, it also leads to banking instability, c.f. the subprime mortgage crisis).

So long as China can rely on the unremunerated labour of women, children, the elderly, and the unemployed in order to feed, clothe, educate, and care for its working class, wages will remain low, exports will grow, and the cycle can continue. But at some point, the costs of reproducing the world's fastest-growing labour force are bound to exceed the abilities of Chinese peasants and the urban lumpen proletariat. And when that day comes, China may well wish that it had invested more in conveyor belts and less in American real estate and tech stocks.

All of this casts doubt on Arrighi's thesis that China has the potential to 'open up for itself and the world an ecologically sustainable development path.'[31] Ironically, it is Arrighi's failure to consider the 'systematic' aspects of Chinese growth – its dependence on exports and American credit, on the one hand, and the non-reproduction of its working class on the other – that blinds him to the less-than-miraculous nature of labour-intensive production. In short, Arrighi doesn't bother to prove that China is experiencing an economic renaissance before advocating a headlong return to the dark ages.

### Info

Giovanni Arrighi, *Adam Smith in Beijing: Lineages of the Twenty-First Century*, Verso, 2007

### Footnotes

**1** Robert Brenner, 'The Origins of Capitalist Development: A Critique of Neo-Smithian Marxism', *New Left Review*, Number 104, London, 1977, pp.25-92.

**2** Immanuel Wallerstein, *The Modern World System*, New York, 1974. Note that Brenner treats only the first volume in the series. When his article appeared, Wallerstein had not yet published the second and third volumes of *The Modern World System*. Brenner also discusses a number of Wallerstein's essays, cited below.

**3** Immanuel Wallerstein, 'The Rise and Future Demise of the World Capitalist System: Concepts for Comparative Analysis', *Comparative Studies in Society and History*, XVI, January 1974, p.391. Quoted in Brenner, op. cit., p.54.

**4** Wallerstein, ibid., p.398. Quoted in Brenner, op. cit., p.32.
**5** Brenner, op. cit., p.32.
**6** Ibid., p.37.
**7** Ibid.
**8** Ibid.
**9** Ibid., p.50. The Marx passage comes from the *Grundrisse*, London, 1973, pp.509-510.
**10** Ibid., p.32.
**11** Giovanni Arrighi, *Adam Smith in Beijing: Lineages of the Twenty-First Century*, London: Verso, 2007.
**12** Mark Elvin, *The Pattern of the Chinese Past*, Stanford, CA: Stanford University Press, 1973, p.314. Quoted in Arrighi, op. cit., p.330.
**13** Arrighi, op. cit., p.32.
**14** Ibid., p.33.
**15** Ibid.
**16** Ibid., p.34.
**17** Philip C.C. Huang, 'Development or Involution in Eighteenth-Century Britain and China?', *The Journal of Asian Studies*, Volume 2, Issue 61, 2002, p.514. Cited in Arrighi, *Adam Smith*, p.39.
**18** Ibid., p.39.
**19** Robert Brenner and Christopher Isett, 'England's Divergence from China's Yangtzi Delta', *The Journal of Asian Studies*, Volume 2, Issue 61, 2002, p.613. Quoted in Arrighi, *Adam Smith*, p.29.
**20** Arrighi, op. cit. p.30.
**21** Ibid., p.39.
**22** Ibid., p.37.
**23** Ibid., pp.365-366.
**24** Ibid., p.366.
**25** Ibid., p.365.
**26** Ibid.
**27** Karl Marx, *Capital: A Critique of Political Economy*, Vol. 1, translated by Ben Fowkes, New York: Vintage Books, 1977, pp.436-437.
**28** John Henry is an American folk hero of the railroads who, according to the legend, outworked the steam hammer (and died) to save his men's jobs, http://en.wikipedia.org/wiki/John_Henry_(folklore)
**29** Arrighi, op. cit., p.8.
**30** 'No Place to Call Home', *The Economist*, 7 June, 2007.
**31** Ibid., p.389.

Daniel Berchenko <dberchenko@gmail.com> is a writer living in New York

# MEXICAN WAVE

Since the 2006 Oaxaca revolt, State repression in Mexico has contributed to popular feeling that peaceful protest has failed. Today, the country is on the threshold of a cycle of armed anti-capitalist struggle, argues Mihalis Mentinis

The repression of the protests in Oaxaca and Atenco in 2006 left people with a strong feeling of 'unfinished business' with the State. A placard in the streets of Oaxaca expressing this feeling showed an enraged Emiliano Zapata holding a gun in each hand: *Nos Vemos en 2010 Cabrones* – 'See you in 2010, bastards'. It is said that there is something like a 'hundred year cycle' in Mexico with the tenth year of every century marking the beginning of a sustained revolutionary effervescence: in 1810, the war of independence against Spanish colonial authority; in 1910, the Mexican Revolution against the dictatorship of Porfirio Díaz; and 2010, an important year of change for the Mayan calendar, seems to meet all the conditions for a new Mexican Revolution – this time against capitalism. Within this atmosphere of anticipation (and preparation), the prospect of a civil war or a revolution dominates the horizon of the future, and Subcomandante Marcos has also warned: 'we are anticipating a great uprising or a civil war.'[1]

The Zapatista revolt in 1994 managed to irreversibly open Pandora's Box for Mexico. Although many were rather premature at that time in celebrating the revolt as the 'New Mexican Revolution', Marcos' rectifying statement that it was a revolution (with a small 'r') which made Revolution possible was on the right track. The revolt unleashed proletarian forces and had a catalytic effect for the formation of new and reactivation of existing indigenous and *campesino* (peasant) organisations, as well as politico-military groups, which have now become forces to reckon with in Mexican politics. After the State's unmitigated repression of popular protests in Atenco and Oaxaca, the politico-military groups in particular have grown much stronger than before, and their belligerence against the State and capitalism is attracting more and more members and supporters from the growing number of people who have lost hope that anything essential can ever be changed by peaceful means. Mexico is on the threshold of a new cycle of struggle that resumes armed and combative actions as indispensable for any fundamental social and political change.

In March of this year in the mountains of Guerrero, some dozens of *campesinos*, some of them representatives of groups and organisations, met in order to announce their integration into the ranks of the 'Insurgent People's Revolutionary Army' (ERPI). The ERPI announced the formation of new

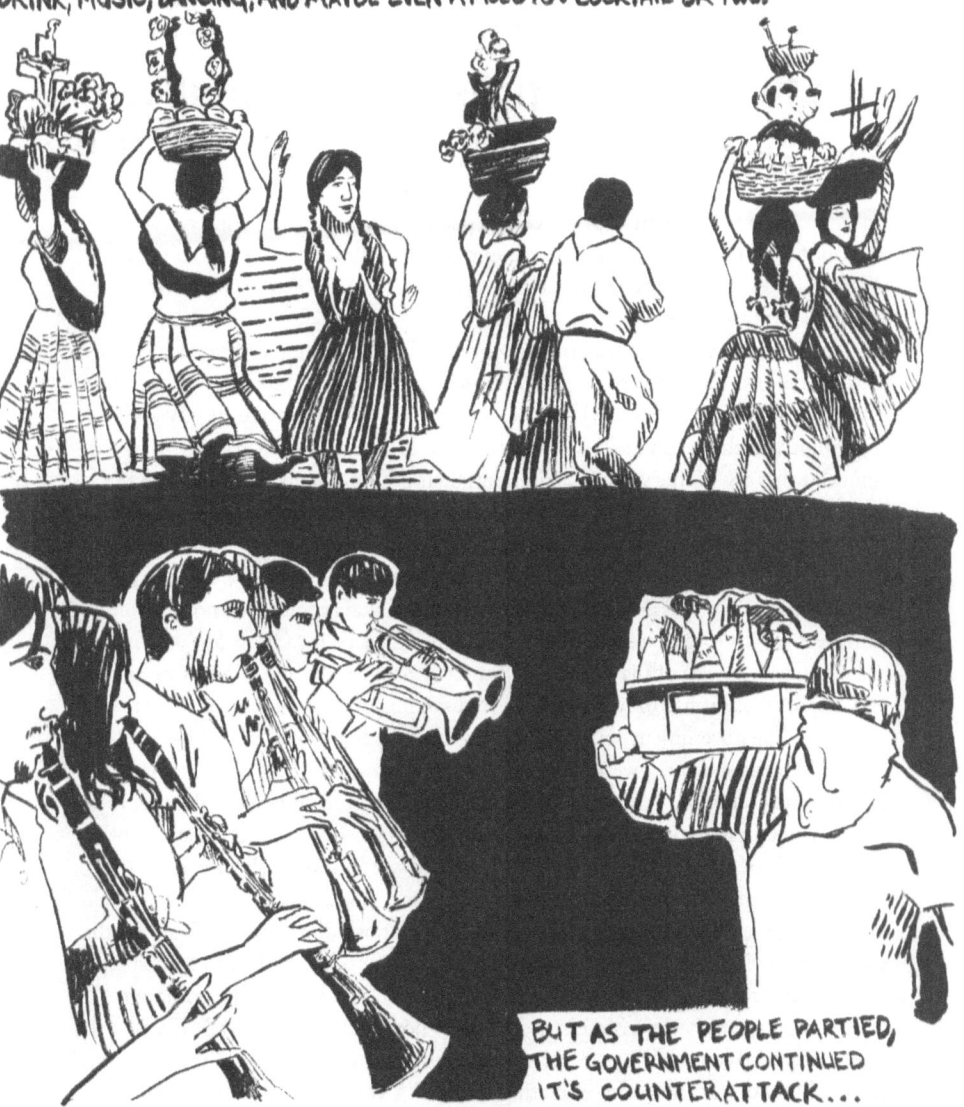

Images: Ana Nimo, from ¡Fuera Ulises!: A Graphic Account From Oaxaca, 2007. The full comic can be downloaded from http://www.metamute.org/en/fuera_ulises_graphic_interpretation_of_the_events_in_oaxaca

columns composed exclusively of indigenous rebels and declared their intention to engage in combative action in order to fight capitalism, defend Mexican oil and natural resources, and fight for the indigenous rights which are circumvented and violated by the repressive government of Felipe Calderón. The participants in the meeting declared their total lack of trust in any political parties and expressed their conviction that the solutions to their problems must pass through the armed

# In Mexico, the prospect of civil war or revolution dominates the horizon

struggle. As a participant explained, 'the pacific way has never worked, exactly the opposite, day after day things become worse'. 'We are not going to wait till 2010 in order to start the revolution in Guerrero,' the ERPI declared, 'we have already started it here.'[2]

Last year, another politico-military group, the 'Revolutionary Popular Army' (EPR), claimed credit for two series of bombings of various pipelines of the State-owned oil company (PEMEX) in various states of central Mexico in July and September. In the second series of attacks in Veracruz and Tlaxcala, the group announced that its 'military units' had undertaken the attacks in order to force the government to hand over two of its militants who had disappeared that same year after being arrested in Oaxaca. One bomb was discovered intact with a message attached: 'Alive you took them, alive we want them back' – an apparent reference to the missing militants. The EPR, however, has a national agenda too and the bombings were meant to deal a strong blow to the president Felipe Calderón (who made security a centrepiece of his presidency), and the neoliberal plans of his government. Although the EPR has only a couple of thousand insurgents and no capacity to overthrow the government militarily, the 'prolonged people's war' that it has declared against 'the anti-people government' has the capacity to strike the State at vital points. The attacks in Veracruz and Tlaxcala resulted in a 25 percent drop in the supply of natural gas available to consumers across Mexico, caused hundreds of millions of dollars in production losses for PEMEX and seriously affected the private sector. Some dozen major companies including Honda Motor Co., Kellogg Co., and Volkswagen had to suspend or scale back operations due to gas shortage. In the case of the Volkswagen plant in Puebla, for example, production had to be suspended for four days resulting in the loss of production of 7,200 vehicles. The bombings were a warning to the government as well as to potential buyers of PEMEX that the intended privatisation of the company would not be an easy process. Given the virtual impossibility of guarding and protecting PEMEX's vast fuel distribution network, the EPR has managed to make a very clear point and introduce itself as a force to be taken seriously in Mexican politics.

Two years ago, in November 2006, a front of five politico-military groups with a presence in several Mexican states took responsibility for bombs planted in the electoral tribunal, the headquarters of the Revolutionary Institutional Party (PRI), and a bank branch. In their communiqué, the rebels declared that these actions would continue against national and multinational companies and State institutions responsible for the 'neoliberal institutional violence' undertaken against Mexican people. They demanded, among other things, the resignation of the PRI-backed and fraudulently elected governor of Oaxaca, Ulises Ruiz, the withdrawal of the 'federal forces of occupation' from the region, and the immediate release of those arrested or disappeared in Atenco and Oaxaca.[3]

These are only a few small instances of a new cycle of armed confrontation in which attacks on prisons, military headquarters and army convoys, sabotage operations against State institutions, but also peaceful demonstrations, protests, occupations and sit-ins have become more and more frequent. In a country where political disappearances, torture, rape and murder of activists is an everyday phenomenon, and where corruption guarantees almost complete impunity for corrupt State officials and paramilitaries, groups like those mentioned above are nurtured and supported by an increasing number of impoverished *campesinos* with no trust in institutional politics.

An important characteristic of the new cycle of struggle is the radicalisation of sectors of the proletariat, both in the cities and rural areas, that one way or another previously maintained some kind of alliance with or faith in the institutional 'Left'. These sectors have now broken from the leftist Party of Democratic Revolution (PRD), and reject it as another mechanism of the State. One of the most recent reasons, albeit not the only one, is that in various Mexican states local bosses have restructured their power relations by switching alliances and linking themselves with the PRD. As a result of this, in places where the PRD has taken power it has tended to maintain the same anti-popular policies and repressive methods as the other parties. In April 2004, for example, in Zinacantán municipality of Chiapas, the PRD authorities commanded an armed attack against a peaceful demonstration by Zapatista supporters demanding access to drinking water. The attack left many wounded and hundreds of refugees, and resulted in a total break between the Zapatistas and the PRD. In a similar case in the state of Guerrero, the PRD government of Zeferino Torreblanca (a businessman) responded with violence to a demonstration against the construction of a hydroelectric megaproject. The lukewarm and often indifferent reaction of the national PRD leadership to all this has further increased the tension, and has greatly disillusioned people.[4]

The break with the PRD was intensified and given a national dimension during the Zapatista *otra campaña* (the 'Other Campaign') in 2006, during the run up to the national election, when Marcos fiercely criticised the PRD and its candidate López Obrador for selling out to financial interests and making alliances with capitalists, calling Obrador the 'left arm of the Right'. The betrayal of the Popular Assembly of the People of Oaxaca (APPO) by the PRD the same year and its clear alliance with and support for the political class, disillusioned many more. Many of the APPO participants, for example, although they insist on a struggle by peaceful means, have completely broken from the PRD and reject electoral politics altogether.[5] The confrontation with Obrador and PRD has forced many of the supporters/allies that the PRD and the Zapatistas shared in common to take sides and decide what kind of

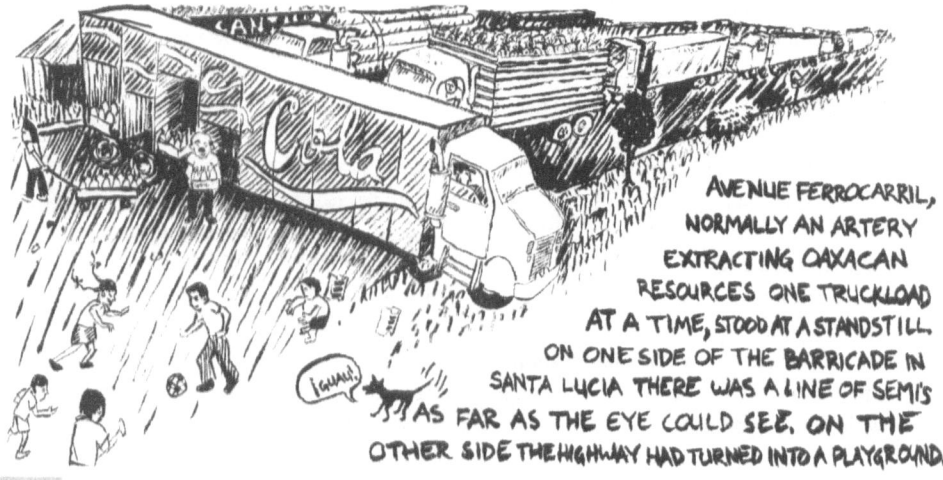

AVENUE FERROCARRIL, NORMALLY AN ARTERY EXTRACTING OAXACAN RESOURCES ONE TRUCKLOAD AT A TIME, STOOD AT A STANDSTILL. ON ONE SIDE OF THE BARRICADE IN SANTA LUCIA THERE WAS A LINE OF SEMI'S AS FAR AS THE EYE COULD SEE. ON THE OTHER SIDE THE HIGHWAY HAD TURNED INTO A PLAYGROUND.

politics they opt for. The Other Campaign has indeed been an important motor of radicalisation of sectors of the Mexican proletariat, and its polemic has brought latent conflicts and antagonisms to the surface. The main contribution of the Other Campaign, however, has been the gradual development of a common anti-capitalist ground that has brought rural indigenous and *campesino* organisations, urban movements, independent labour unions and various socialist groups closer together. The Zapatista framing of indigenous demands within an anti-neoliberal agenda, and their recent use of the term 'capitalism' – previously absent in the discourse of the movement, has been an important factor in this direction.[6] It is indicative, for example, that one of the easiest decisions reached by the popular assembly in Oaxaca was precisely the one related to the need to give a clear anti-capitalist orientation to their struggle.[7]

If the Other Campaign has been significant in inciting proletarian fight-back, Oaxaca is the name of the event (in its Badiouean sense of the term as a rupture with 'what is') that marks the new cycle of struggle. The Oaxaca commune, as it is now commonly referred to, opened up different possibilities and produced the conditions for a new dynamic cycle of struggle, building on and at the same time expanding the

## The Oaxaca commune produced the conditions for a new cycle of struggle

cycle of struggle initiated by the Zapatista revolt. In a period in which right wingers and leftist academics were talking about the decline of the Zapatista project and the virtual impossibility of non-taking-power politics, (even mockingly dubbing Marcos the 'subcomediante' - e.g. Žižek), the commune of Oaxaca showed that radical direct democracy practised in small indigenous communities could be taken up on a larger scale, on a city, level with the participation of diverse ethnic and socio-economic sectors, and render the government completely redundant. For indeed the commune of Oaxaca rendered completely redundant the local and national government for more than six months.[8] The APPO can be described as what Shaj-Shuja calls a 'Zone of Proletarian Development' (ZPD): a space in which various sections of the proletariat (e.g. *campesinos*, indigenous, teachers etc.) learn from each other and radicalise their consciousness in the process of joint activity.[9] Its consequences and effects are still to be seen.

The importance of the Oaxaca commune lies in the fact that it has become the symbol of a successful people's government, and therefore a model for prefiguring the future, for imagining how a revolutionary outbreak could be. Marcos too seems to see the future of struggle in Mexico as taking the form of local uprisings and communes. When he was asked about who would lead the Revolution or civil war he talked about, he responded: 'people themselves, everybody in the place where they live, and all in a network of mutual support', and he gave the example of Oaxaca where 'there are no leaders and no caudillos; it is people who have organised themselves.' 'Like this it is going to be in the whole country,' he said.[10] The violent repression of the APPO then aimed, first and foremost, to prevent the example from spreading, given that people in other areas of the country had already started organising themselves in popular assemblies. The fact that APPO had defined its ultimate goal as an autonomous form of people's government for the entire state was also key to the ensuing repression. In this atmosphere of anticipation and preparation for the future, the second round of the Other Campaign undertaken by the Zapatistas in March 2007 had as its main objectives to strengthen further the common anti-capitalist basis of the various rural and urban movements across Mexico, to bring together proposals and demands from various local struggles into a 'national programme of struggle' and further strengthen solidarity and support networks in the conditions created by the newly repressive administration.

What are these conditions? Since the beginning of the Fox administration in 2000, it had become clear that Mexico was interested in strengthening ties with the US and moving away from other Latin American countries. The public rupture with Fidel Castro in the Monterrey 'poverty summit' in 2001, the Mexican candidate's cooperation with the US against the candidates of several Latin American countries for the position of the Secretary General of the OAS ('Organisation of American States'), and the rupture with Hugo Chávez etc., all attested to Mexico's US oriented future. This agenda has now been taken up by Calderón who is interested in renewing a partnership and strengthening further ties with the US.[11] Given this, it is unlikely that the government would ever risk upsetting the US by bringing to the table even one of the most basic popular demands brought up during the tortilla riots in 2007: the revision of the NAFTA's ('North American Free Trade Agreement') section on agriculture in relation to maize (the basic ingredient of the tortilla) and beans (products that enter Mexico from the US with no restrictions and are normally genetically modified threatening *campesino* selling power and corn/bean diversity respectively).

## The Zapatista revolt in 1994 managed to irreversibly open Pandora's Box for Mexico

Calderón has instead opted for repression and the militarisation of the country in order to control proletarian forces. With the pretext of combating drug trafficking, check points have been set up everywhere and large army units have been dispatched to areas perceived as prone to insurrection and upheaval (e.g. Chiapas, Oaxaca, Guerrero etc.). The militarisation of the country obeys the new security dogma of Mexico; a dogma that implies the increasingly vital participation of the US in terms of strategic guidance, training of the military and the police, supply of military technology and so on. All this is accompanied by the installation in Mexico of mercenaries, private foreign security companies and the expansion and further arming of paramilitary groups. For it is certain that the increase in arms as a consequence of militarisation will result, sooner or later, in a

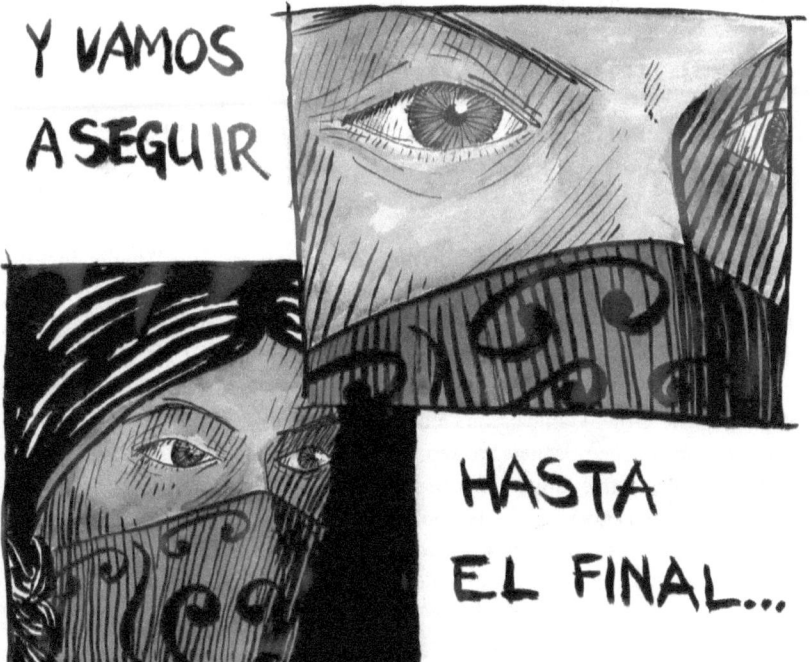

"NOW US WOMEN HAVE RISEN UP, AND WE WILL CONTINUE UNTIL THE END..."
—ANONYMOUS PARTICIPANT IN THE OCCUPATION OF CANAL 9

number of these arms ending up in the hands of paramilitary groups. This is the Mexican version of the 'plan Colombia' intended to intimidate, repress popular protests, imprison, torture and murder activists, perform anti-guerrilla operations and train paramilitary groups under the guise of the anti-cartel war. However, so far everything seems to suggest that Mexico's proletarians are far from being pacified and controlled.[12]

It is difficult to say what the future holds for Mexico. Although there is definitely no clear plan to move forward, it is the first time that armed and non-armed organisations and groups have come so close in their agendas. Back in 1996, when the Zapatistas enjoyed a rosy romance with the middle classes, they had arrogantly rejected the EPR's solidarity and support: 'we don't want your support, we don't need it, we don't seek it'.[13] Some years later, however, during *la Marcha* in 2001, the Zapatistas revised their position towards armed revolutionary groups, among them the ERPI and EPR, recognising their roots in Mexican social and political reality. Despite the fact that in 2005, just before setting off for the Other Campaign, the

## APPO defined its goal as autonomous people's government for the entire state

EZLN ('Zapatista Army for National Liberation') made clear that it still maintains its commitment to the path of political struggle through peaceful initiatives, and announced it would not establish any kind of secret relations with politico-military organisations, it is obvious that there is a rapprochement between the EZLN and other armed groups in terms of solidarity and mutual respect.[14] Marcos' gestures of solidarity during the Other Campaign included a visit to a prison near the capital of Oaxaca where many of the prisoners are kept for their alleged links with the EPR, and the unconditional active support that the Zapatistas offered the APPO (APPO counted on participants and tactical – but not military – support from several of the region's armed groups). For their parts, armed groups like the EPRI have spoken positively of the Zapatistas, implying an understanding of the importance of the non-military political strategies of the latter:

> we see positively the work of Subcomandante Marcos with the Zapatista Army for National Liberation in Chiapas... [T]heir work is well structured and their actions give them results'.[15]

The EPR has also announced that they will not engage in any acts that could jeopardise the EZLN. In the new cycle of struggle, differences among various armed and non-armed organisations have been left aside to a great extent, and emphasis is instead being put on the anti-capitalist and anti-state demands, hopes and desires they all share in common. In Mexico, proletarians have made a combative come back and we can aver with certainty that 'all this has just begun'.[16]

## Footnotes

**1** Hermann Bellinghausen: *Marcos*: 'Estamos en vísperas de un gran alzamiento o una guerra civil', *La Jornada*, 24 November 2006, http://www.jornada.unam.mx/2006/11/24/index.php?section=politica&article=015n1pol

**2** Sergio Ocampo Arista: 'en respuesta a los ataques del gobierno, indígenas se suman a la lucha del ERPI', *La Jornada*, 25 March 2008, http://www.jornada.unam.mx/2008/03/25/index.php?section=politica&article=012n1pol

**3** Cinco organizaciones se adjudican los bombazos, *La Jornada*, Tuesday, 7 November 2006 http://www.jornada.unam.mx/2006/11/07/index.php?section=politica&article=007n1pol

**4** Rosalva Aída Hernández Castillo, 'The Indigenous Movement in Mexico: Between Electoral Politics and Local Resistance', *Latin American Perspectives*, 2006, 33: pp. 115-131.

**5** Luis Hernández Navarro, 'APPO, PRD y elecciones en Oaxaca', *La Jornada*, 22 March 2007, http://www.jornada.unam.mx/2007/05/22/index.php?section=opinion&article=019a1pol

**6** Mariana Mora, 'Zapatista Anti-Capitalist Politics and the Other Campaign: Learning from the Struggle for Indigenous Rights and Autonomy', *Latin American Perspective*, 2007, 34: pp. 64-77.

**7** Gustavo Esteva: 'The Asamblea Popular de los Pueblos de Oaxaca: A Chronicle of Radical Democracy',. *Latin American Perspective*, 2007, 34: pp. 129-144.

**8** The 'subcomediante' pun was coined for the first time back in 1996 by the anti-Zapatista right-wing press in Mexico in order to deride Marcos, and it was employed several times since then by anti-Zapatista media. Žižek has taken up the term, misattributing its origin to Mexican leftists, and employs it in order to convey what he sees as the lack of realism and naïveté of non-taking-power politics. See: http://www.lrb.co.uk/v29/n22/zize01_.html

**9** Masteneh Shah-Shuja, *Zones of Proletarian Development*, London: OpenMute, 2008.

**10** Hermann Bellinghausen, 2006, http://www.jornada.unam.mx/2006/11/24/index.php?section=politica&article=015n1pol

**11** See: Jan Rus and Miguel Tinker Salas, 'Introduction: Mexico 2006-2012: High Stakes, Daunting Challenges', *Latin American Perspective*, 2006, 33: pp. 5-15.

**12** While writing this article, the EPR has sent out another communiqué refusing direct dialogue with the 'criminal government' and calling for a struggle with all means against the privatisation of PEMEX. See: Alonso Urrutia, 'El EPR reivindica la lucha armada y rechaza una rendición incondicional'. *La Jornada*, 13 May 2008, http://www.jornada.unam.mx/2008/05/13/index.php?section=politica&article=011n1pol

**13** EZLN, 'To the Soldiers and Commanders of the Popular Revolutionary Army', 1996, http://flag.blackened.net/revolt/mexico/ezln/ezln_epr_se96.html

**14** EZLN, 'Sixth Declaration of the Lacandon Jungle', 2005, http://www.anarkismo.net/newswire.php?story_id=805

**15** Sergio Ocampo Arista, 2008.

**16** 'All this has just begun' was EPR's warning to the government after the second bombings of PEMEX installations in 2007.

---

Mihalis Mentinis <m.mentinisATgooglemail.com> is a member of the Discourse Unit at Manchester Metropolitan University

# ANY OTHER BUT OUR SELVES

Contemporary curators are loving the alien, the sacred and the cultic. But far from challenging contemporary social mores, this Other-worship is just an orthodox postmodern denigration of human agency, argues J.J. Charlesworth

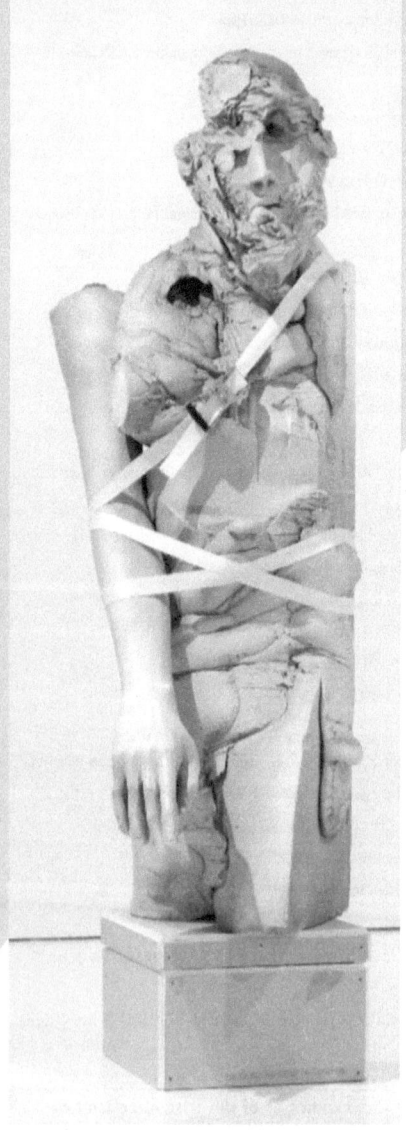

Image: Matthew Monahan, *Phantom Limb*, 2008

It might have been coincidental that this year two big international group shows should look to Mars for their curatorial pretext. In London, the Barbican's spring show The Martian Museum of Terrestrial Art opened only a day after the latest incarnation of the Carnegie International – at Pittsburgh's Carnegie Museum of Art – which was subtitled Life on Mars. The two shows proposed superficially different uses of the metaphor of extraterrestrial life to reflect on contemporary artistic practice, yet underlying these apparent differences, one could detect common themes that are now strongly influential in Western culture – that of a pessimistic apprehension of impending disaster; a profound sense of uncertainty and disorientation regarding human society's claim to progressive agency; and a kind of post-historical estrangement from the experience of modernity. What drives these, however, is a now common theoretical and political celebration of what might be called the Absolute Other. Along with the more recently opened mega-show After Nature at New York's New Museum, these shows all share a peculiar critical operation – the use of a strategy of displacement of the human subject from which to 'look back' or 'look from afar' on human life. Or, as contemporary theoretical jargon would have it, the effect of transposition that results from the attempt to occupy the position of 'the Other'. In their attachment to the Other, these shows reveal a common distrust of, and estrangement from, that much maligned humanist concept, the Subject. It's an assimilation of earlier post-structuralist critiques which manifests itself as a form of extreme scepticism regarding the legitimacy of human subjectivity – psychological, cultural and social – and as a consequence, it is the legitimacy of human agency *in toto* that is effectively called into question.

But while these big offerings tend to share a very liberal sense of melancholia, other more apparently radical positions work over similar ground. In London, Pil and Galia Kollectiv's show The Institute of Psychoplasmics, at Pumphouse Gallery, proposed a more theoretically explicit use of the Subject-Other division, in which the notion of the 'cult' is mobilised as absolute Other, as a critique of a Subject synonymous with Capitalism. Nevertheless, whether appealing to mainstream culture's misanthropic liberal anxieties, or claiming a site of monadic exclusion from the Subject, these recent curatorial positions propose the Subject and human subjectivity as something to be escaped, denied and criticised from *outside*. Yet this throws up a paradox; who is this absolute Other who claims alterity, if not another, still-present Subject?

In Martian Museum, the critique of the Subject was conducted on the terms already long-established by the post-colonial critique of the Other; here, Martians were substituted for the position of the ethnocentric Western Subject; gazing on the works of Western contemporary art as if these were the artefacts that Western anthropology has long designated as the products of the 'primitive' Other. As curator Francesco Manacorda confidently declares in his introduction to the Martian Museum's catalogue:

> The Martian perspective allows for a reassessment of the art object from an alien standpoint: thus mimicking the way that Western anthropologists historically interpreted non-Western cultures through foreign eyes. Looking at art as though from outer space offers the potential to make the familiar strange and to turn the dominant Euro-American art tradition into the 'other'.[1]

However, the explicit political motive that underpins why one should want to perform such an act of 'othering' is spelt out by the Barbican's directors in their preface:

Any Other But Our Selves

In parodying the way that Western anthropologists, ethnographers, and art historians have historically viewed non-Western culture through alien eyes, [Martian Museum] questions the hubris of any culture's pretence to fully understand another, through the acquisition of objects and their didactic display in a museum.[2]

The notion of Otherness is here presented as an insuperable barrier to the understanding of other cultures, as well as typifying the 'hubristic' hegemony of the Western gaze, which is why it is easy for the writers to turn the gaze into something 'alien'. It may be more than a metaphor running away with itself, but to suggest that an 'ethnocentric' Western gaze might see the 'Other' as 'alien' betrays something of how absolute the notion of the Other has now become in common discussions of cultural difference: if there were indeed aliens, only their 'otherness' to our humanity could be so absolute.

What is unsettling in this casual use of the 'alien' 'Other' as a (self-) critical strategy against the 'Western' subject is that cultural otherness between human cultures can – in reality – never be so immobile or absolute. The cultural dynamics of hybridity, heterogeneity and globalisation now challenge, more than ever, the Western hegemony that could generate the post-colonial critique of the Other in

## the legitimacy of human agency in toto is effectively called into question

the first place. And the funny thing about Martian Museum is that the fictional Martian curators seem only to have looked for art in Europe and America. The contemporary reality of art's now rapidly-globalising modernity is nowhere to be found on their sensor instruments – Martian Museum is distinct in almost entirely excluding art from the modernising, industrial Far East.

Martian Museum is of course not so original. The critical excavation of the fantasy of the alien-as-Other, hooked up to a contemporary critique of ethnic difference, had already been made by the InIVA exhibition Alien Nation at the ICA in late 2006, a show which presented the work of artists who 'adopted the figure of the extraterrestrial and the alien(ated) landscape'.[3] If the reified and static difference favoured by liberal multiculturalism could be coded in contemporary art via the figure of the extraterrestrial in Alien Nation, then in Martian Museum it extends further to a decentering repudiation of, and anxiety about, the identity and subject-position of 'the West'. To be a Western subject these days means, it would seem, continuously disaffirming one's Westerness, only to reassert it in negative form, while perpetuating the division of cultural Self and Other.

This 'Othering', presented as a sort of therapeutic critical device, seeks to problematise the secure subject position of our supposedly Western ethnocentrism. Couched in the language of contemporary post-colonialist and multiculturalist discourses of difference and the Other, these are conventional and orthodox rehearsals

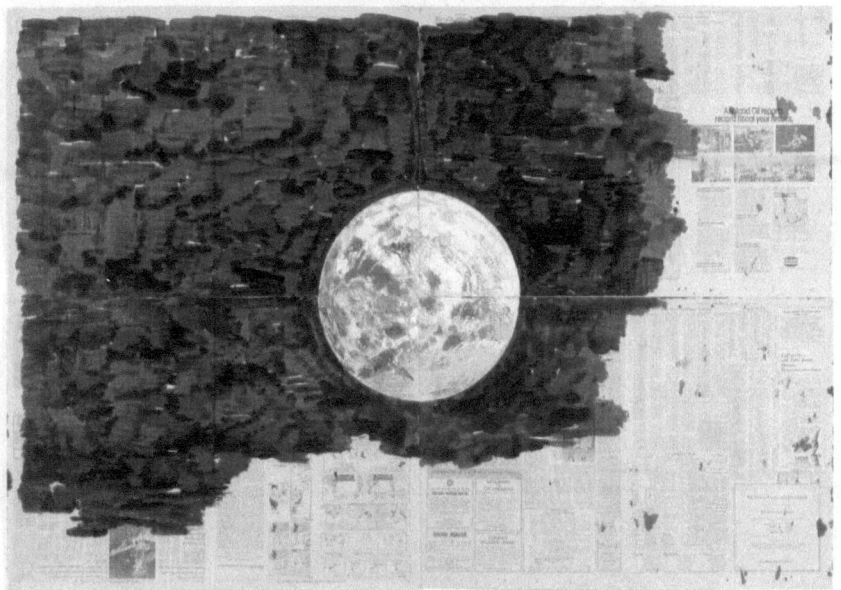

Image: Paul Thek, *Untitled (Earth Drawing I)*, 1974

of the accepted politics of cultural difference. But the insistence of seeing the Self from the perspective of the Other echoes beyond questions of cultural difference, into one of a broader estrangement with regard to contemporary human experience, and a broader 'alienation from' – for want of a better term – the Subject. Pointedly, in his essay to the catalogue accompanying Life on Mars, curator Douglas Fogle refers to Paul Thek's 1974 painting *Untitled (Earth Drawing I)*, an image of the Earth seen from space, and asks questions of peculiarly existential intensity: 'Are we alone in the universe? Do aliens exist? Or are we, ourselves, the strangers in our own worlds?' Musing on the experience of watching reports of the plight of the survivors of Hurricane Katrina, signalling for help from their rooftops 'as if attempting to communicate with an alien culture without the aid of a universal language translator', Fogle goes on to declare that, 'If the events of the last few

Any Other But Our Selves

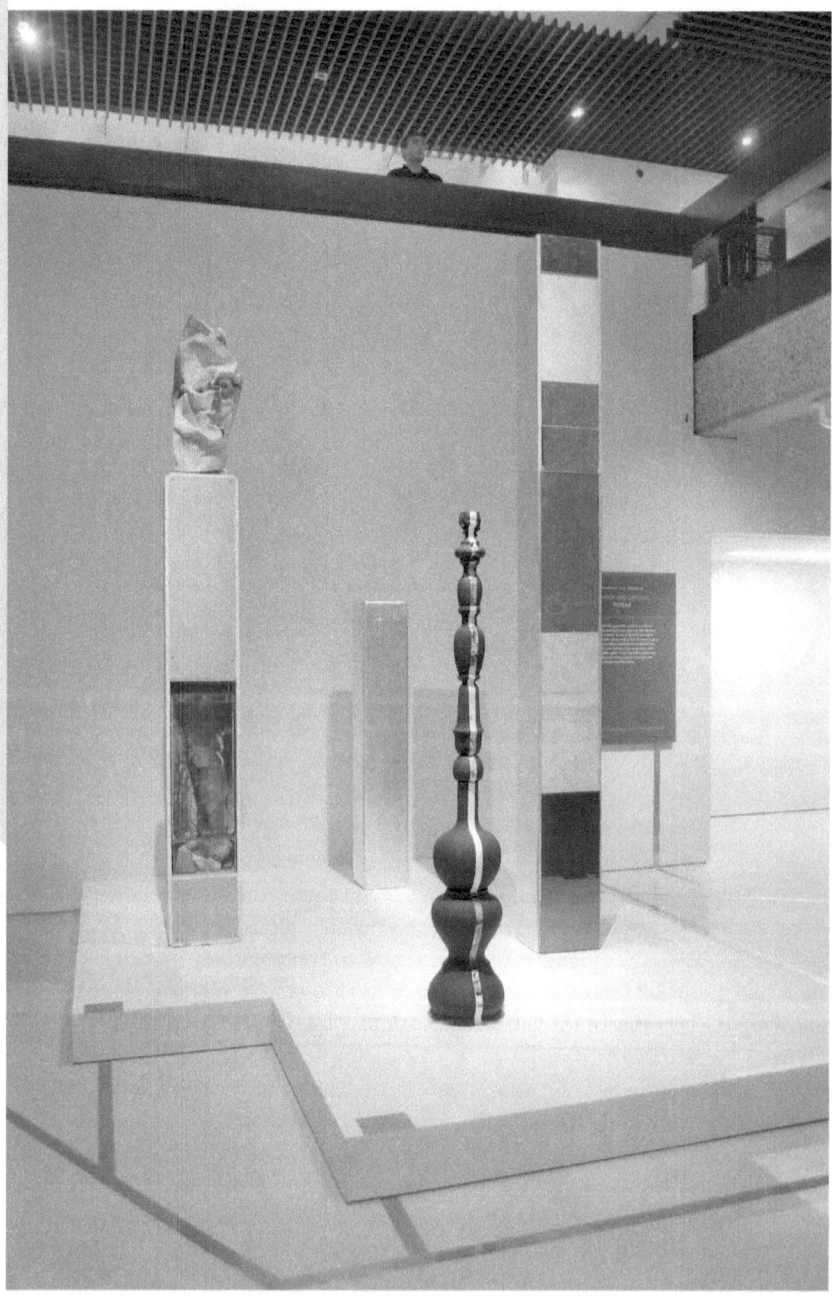

Image: 'Totems' section of The Martian Museum of Terrestrial Art

years have taught us anything, it is that many of us are indeed strangers in our own worlds.'[4] The notion of self-estrangement in place, Fogle continues with reference to that other cultural projection of subjective Otherness, the Zombie, rallying Samuel Beckett and Cormack McCarthy to a 'meditation on the hopefulness of

# globalisation now challenges the Western hegemony that generated the post-colonial critique of the Other in the first place

humanity in the face of total despair and utter devastation'. 'Like the characters in Beckett's play,' Fogle dismally declares, 'we find ourselves waiting as well, for ours is a culture of zombies in which we are slowly and at times deliberately eating ourselves to death.'[5]

That a curator of a major international show could express such a wild thesis on the state of contemporary society says something about how acceptable such ultra-subjective, pessimistic observations are to current cultural and political anxieties towards a world in which events appear out of our control. Fogle's use of language is itself revealing of a sense of passive, helpless – essentially agentless – acceptance, in which events happen to us, while we are also simultaneously the perpetrators – yet the hyperbole is fixed on an altogether prosaic, real-world experience:

> The question 'Is there life on Mars?' is a rhetorical one posed in the face of an increasingly accelerating world, in which global events, at once political, social, natural, and economic, seem to challenge and threaten to overtake our most basic forms of everyday existence.[6]

Impending disaster, it seems, is only a curatorial flourish away. With Life on Mars, the thematic of disaster is bound up to a profound anxiety about the disorientation of the contemporary subject faced with uncontrollable change. Suspended metaphorically between Mars and Earth – the subject of Life on Mars becomes a 'metaphorical quest to explore what it means to be human in this radically unmoored world'.

Why the world should be any more radically unmoored than at other times in history is never made clear. But if in Life on Mars the dislocation of the contemporary subject is couched in a spatial metaphor, After Nature generates a similar existential crisis in terms of history, yet again pointing to contemporary political and social reality in terms which are distinctly apocalyptic. Curator Massimiliano Gioni's introductory essay is titled with a quote from Blaise Pascal: 'The collapse of the stellar systems will occur – like creation – in grandiose splendor'; a suitably vast and inhuman terminus to any sense of temporal progression.[7] After Nature takes its title from W.G. Sebald's eponymous poem triptych, and much of its rhetorical inspiration from Werner Herzog's 1992 film *Lessons of Darkness*, and the result is an insistently post-historical, (or post-apocalyptic) vision of human life:

> Many of the artworks and artefacts in this show ... might recall the relics of a lost civilization. Enclosed in vitrines, captured on film, or sketched with charcoal, they could be the mysterious findings of some ethnographer, or the experiments of a scientist desperately trying to save the last remnants of this planet, while documenting its imminent collapse. [...] It is a land of wilderness and ruins that exists in an imaginary time zone suspended between a remote past and a not-so-distant future.[8]

We're back onto the territory of the ethnographer, of the archaeologist and the conservationist who is, implicitly, beyond, or outside, of our own historical moment – the 'Other' to our own present-bound experience. But whether cast as aliens

## the apprehension of a totalising Capitalism leads to a retreat to the cultish and inhuman

from another planet or men from somewhere in the future, these displacements invariably seek to position our contemporary culture as something itself already alien – something that we should in some way treat with self-conscious distance or with distrust. With Martian Museum, it is the distrust of the 'hubristic' Western subject, even when it paradoxically has to shore up that Eurocentric gaze by the deliberate exclusion of the contemporary non-Western artist. In Life on Mars, the destructively displaced and spatialised multiplicity of cultural identities and positions, and the out-of-control character of contemporary experience are counterposed with works of melancholic and pacific reflection on the intangible and inhuman immensity of the cosmos. With After Nature, time and history are both post- and pre- modern. This pre-/postmodernism is an explicit anti-modernity, which, when not obsessed

J.J. Charlesworth

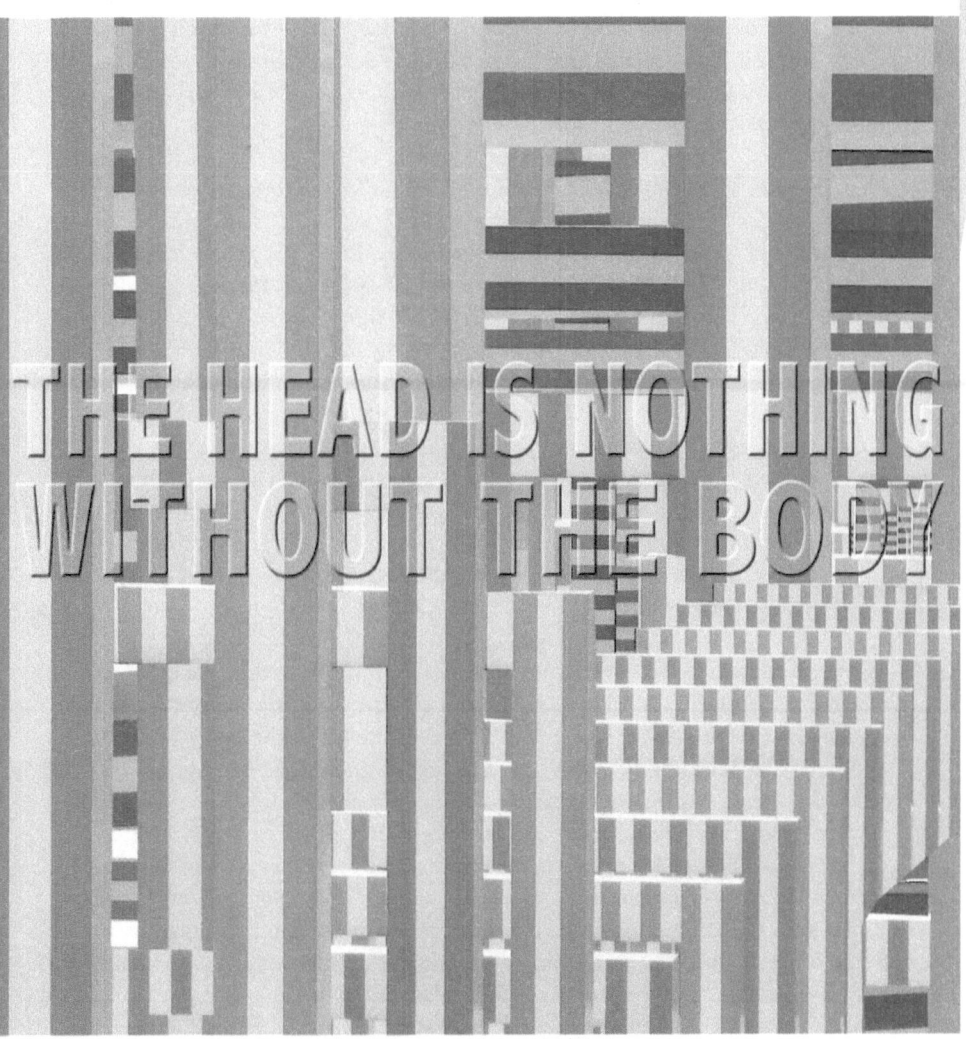

Image: Amanda Beech, *The Head is Nothing without the Body*, Triptych (1 of 3), 2007

with the before/after exclusion of humanity under the sign of ecology, admits human subjectivity only via the mystical, the sacred and the religious: 'Many artists in this exhibition seek a prophetic intensity,' Gioni writes, '... fascinated by mystic apparitions, arcane rites, and spiritual illuminations.' Intoxicated with his theme, he declares that:

> what these artists share is [...] a desire to charge art with a magical power, invest it with contents and forms that are meant to contrast its secularization, and aim at bringing it back into a sphere that is, if not religious, at least sacred or obscure, like a mystery cult.[9]

Here, Martian Museum catches up with After Nature, as they both endeavour to 'depict a humanity caught in cryptic rituals or in adoration of mysterious divinities'.

Cults and cryptic rituals also abound in The Institute of Psychoplasmics, although Pil and Galia Kollectiv mobilise the notion of the cult to more critical ends. Working out from David Cronenberg's 1979 film *The Brood*, the Kollectiv's project approaches the figure of the cult as a strategy to question the problem of political otherness and differentiation within the social body of capitalism:

> If there is no outside to an ever absorbent Capitalist regime, art has no grounds from which to critique it. Religion increasingly seems to offer a way out of the impossible dialectic of outside and inside that Capitalism proposes. Art takes on a similar role, allowing for a limited externality to emerge within a social body whose unlimited expanse is equated with the expansionist project of global Capitalism.[10]

'If there is no outside' is, however, a pretty big 'If'. But it is in this that the figure of the Other – even for would-be radicals such as the Kollektivs – takes on its full contemporary significance. To argue that religion and art might be a way out of a totalising Capitalism would seem a peculiar argument to earlier generations of revolutionaries, for whom capitalism was necessarily historically contingent, and anyway already riven by an internal dialectic in which the Other might have been the agency of class, and the Subject the proletariat. But let's not dwell on bygone Grand Narratives. That one can suggest that there is no outside to Capitalism is a partial symptom of the disappearance of the narratives of the revolutionary Subject. The fall-back to art, and art which mimics the cultish religious refusal of liberalism's totalising community, is a sad shadow of the dynamic of earlier political divisions – the dialectical energy of the revolutionary historical Subject reduced to a frozen, mute abstentionism.

It's also telling that The Institute of Psychoplasmics combines an (almost) ironic affection for the forms of cultish community with a more extreme allegiance to an effectively post-human subject. The reference to *The Brood* is not idle – Cronenberg's film about an extreme psychotherapy group, in which the patients' emotional states become physically expressed, chimes with the various fantasies of a libidinally liberated post-human body – from Diann Bauer's miniaturised figures to

J.J. Charlesworth

Seth Coston's paintings of suited men who erupt with multiple phallus tentacles – in which the Cartesian mind-body division is comprehensively dissolved. 'The Head is Nothing Without the Body' declares a work by Amanda Beech, and human heads are indeed hard to find in The Institute of Psychoplasmics – a fashionable Acephalism that shows how far the post-structuralist critiques of the Subject have become engrained in both liberal and radical thinking. One might be tempted to reply that there is of course 'no-body without a head'. The Institute of Psychoplasmics resorts to the Cult as a fantastical absolute Other to the political Subject of a Capitalism that can no longer be transcended, but it also harbours a more profound – and conventional – desire to be absented from the Human itself.

What lies beneath Martian Museum's superficial critique of the 'Western' subject is really the rejection of the notion of a potentially universalising Subject that transcends the 'difference' of any given cultural subject. In Life on Mars and After Nature, the apprehension that contemporary society has abandoned any claim to progressive historical agency, and any progressive narrative of human history, manifests itself in each show's peculiar characterisation of the experience of history, in which active human agency has been removed. In Life on Mars this appears in the figuring of reality in terms of an anxious and essentially incomprehensible, never-ending present, to which we are helplessly subject. With After Nature, it is contained in the vision of a pre-/post-historical world in which humanity has become extinct, or

# reversals of the Other actually return us to the universal human Subject, but in negative form

What unites these shows is that they are each symptomatic of a complex disquiet about the nature of human experience and human subjectivity in an epoch in which the claims of the humanist, universalising Subject appear exhausted and discredited, and in which those human-centred, progressive narratives of human society and subjectivity have all been dismissed.

at most has retreated to pre-modern, pre-secular and pre-rational forms of subjectivity. And with The Institute of Psychoplasmics, the apprehension of a naturalised, totalising and ahistoric Capitalism leads to a semi-ironic retreat to the cultish and the inhuman, as strategies to escape the problem of an apparently undifferentiated and unchangeable political stasis.

Such curatorial anxieties are neither that unusual nor coincidental. After all, the dissolution of the humanist Subject has been the mainstay of post-structuralist philosophy and theory for a generation. Daniel Birnbaum is no fool when, in his contribution to the Life on Mars catalogue, he variously draws on Deleuze, Foucault and Kojève to point to 'new modes of subjective life'. He ends up in the post-historic and the post-human, of course:

> What is Man and what is Animal today? What are the distinctive features of life after History, and who are we, breathing this posthistorical air, taking advantage of the lightness, and writing the postscript to historical Man?[11]

To tell 'us' who 'we' are, Birnbaum coopts Kojève, in his lecture on Hegel's *Phenomenology of Spirit*:

> The historical process of work and negation, as analysed by Hegel and then, in a different fashion, by Karl Marx, is what turned the animals of the species Homo sapiens into humans, but this process has come to a close. We have reached the end of History yet, it seems, we are still around. Kojève goes on: *The disappearance of Man at the end of History is not a cosmic catastrophe.*[12]

The now-orthodox heresies of post-structuralism's assault on the humanist Subject might be fine for the theory class, but they now combine with the conservative theme of the 'end of History', and the cultural and political consequences are novel and unpredictable. If post-structuralism has been declaring, for the past three decades, that Man is 'over', then it is not surprising that other bugbear of anti-humanism, the dialectic of History, should be declared 'over' too. Yet, it seems, 'we' are 'still' around.

Let's not blame curators for reproducing intellectual trends out of their control. That such big shows express similar perspectives merely reveals that they chime with the zeitgeist, in which the theoretical dismantling of the Subject now starts to intersect with a more general mood of cultural exhaustion and political terminus. The continued motif of the 'Other', and the scrutiny of humanity from its position is necessarily a fantasy, but it signals how estranged from the concept of a centred Subject – psychological, social, historical – we now find ourselves. As James Heartfield argues in his *The Death of the Subject, Explained*:

> Like the concept of the posthuman, the elevation of the Other corresponds to a degradation of the Subject. The desire to relinquish the Self, leave the species, stop persevering in one's being are all essentially the same death-wish.[13]

J.J. Charlesworth

Or as the humanist writer Josie Appleton has recently observed:

> Everywhere there is a discomfort with seeing things from a human point of view, or pursuing human interests. Indeed, it seems that we would rather see things from any point of view but our own, and defend any interests other than our own. This appears not as a craven attitude to the gods, but a craven attitude towards nature ... 'Anthropocentrism' has become a dirty word, spat out along with 'humanocentrism', homocentrism and 'humanism'. Indeed, to see the world from a human point is 'speciesism' ... Instead, theorists hunt around for other loci of value: 'zoocentrism', 'ecocentrism', even 'cosmocentrism'.[14]

One doesn't have to go far to make the connection between the philosophical 'degradation of the Subject' and the pursuit of 'other loci of value' which deny a human point of view, nor are they hidden in these curatorial outings. The point of view of the alien, of the inhuman, of the pre- and post-historical, the mythical, and the ecological become the subjects for a world view in which humanity sees *itself* as other. And among these there is no better, nor more influential, manifestation of this self-estrangement of human subjectivity than the position of environmentalism. One might argue that environmentalism finds itself the true heir of post-humanism and the end of History, given that under its sign, humanity becomes Other to environmentalism's Subject – Gaia, the ecosphere.

The philosophical paradox here is that such reversals of the Other actually return us to the universal human Subject, but in negative form. This Other humanity is seen as a single entity or identity, yet these melancholic dystopias in which humanity is observed as a specimen or a relic are necessarily the projection of human subjects, and we do not 'stop perservering in our being' even though we attempt to theorise ourselves out of existence. But this philosophical paradox goes hand in hand with political terminus: these are not cultural, ethnic, social or political Others, Others that might be susceptible to encounter, dialogue, negotiation, and through that process – the active agency of human politics and history – the overcoming of that Otherness. This is instead an absolute Other, the human seen (by itself) as passive object, rather that active Subject. And it is here that the post-human meets the post-political. If it appears on one hand as if the world is out of our control, that we are the victim of forces we have no influence over, then it also appears that 'we' are unstoppably destroying it on the other – nowhere in this is there a sense of active, purposeful human agency. If this is a consequence of the death of 'Grand Narratives' then the engine of those narratives, the human Subject, is clearly on the verge of ex-

Any Other But Our Selves

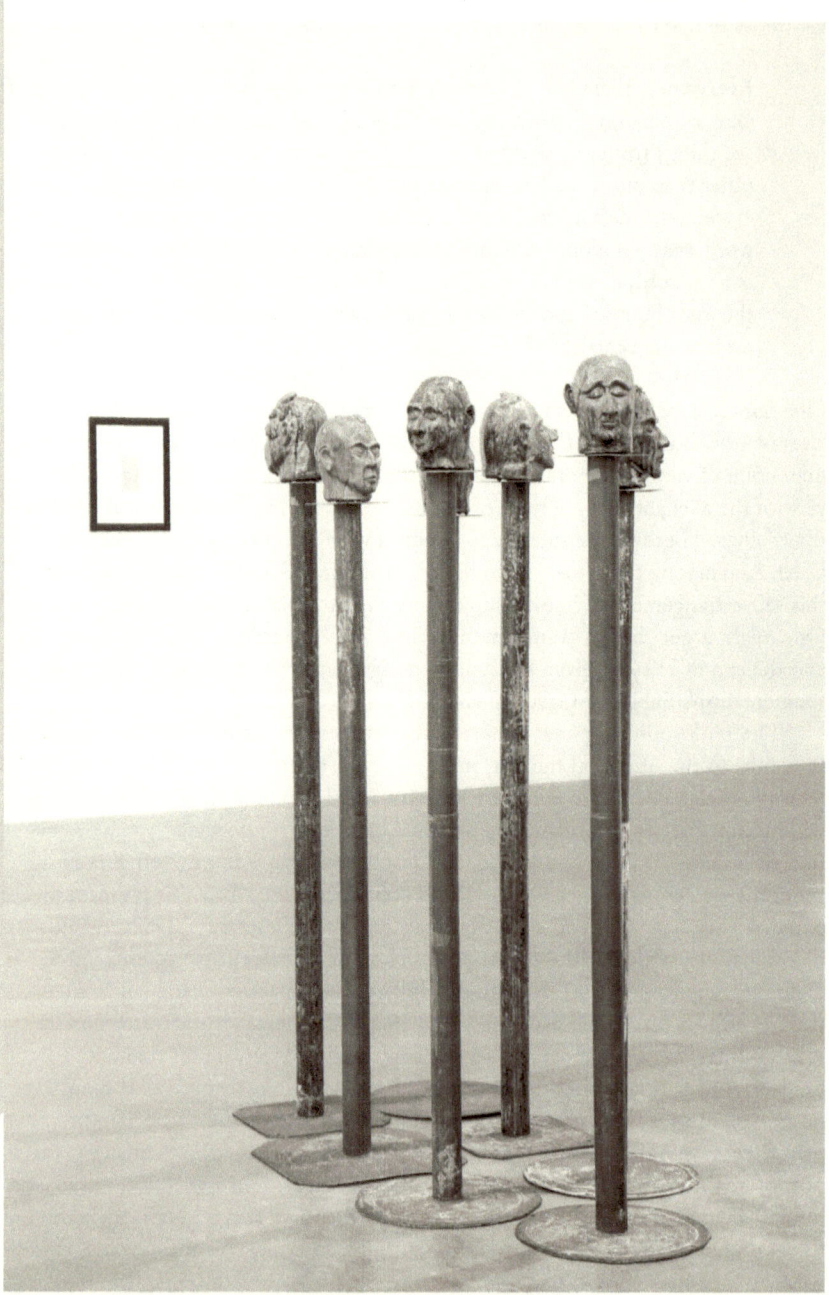

Image: Thomas Schütte, *The Magnificent Seven*, 1993

tinction, and from then on, we have little else to do than to muse about our own post-humanity and wait for the Apocalypse. The cultural melancholia expressed in shows from Martian Museum to After Nature points us to the novel political problem of making sense of what we mean by, and expect of, human subjectivity and human agency; of whether we want to develop new terms on which to remake the human Subject, or instead disappear into the vague hinterlands of the Other.

**Info**

Martian Museum of Terrestrial Art: Barbican Art Gallery, London, 6 March – 18 May 2008

The 55th Carnegie International: Life on Mars, The Carnegie Museum of Art, Pittsburgh, 5 March 2008 – 11 January 2009

The Institute of Psychoplasmics, Pump House Gallery, London, 9 April – 26 May 2008

After Nature, The New Museum, New York, 17 July – 21 September 2008

**Footnotes**

**1** Francesco Manacorda, *Martian Museum of Terrestrial Art*, London and New York: Barbican Art Gallery and Merrell Publishers, 2008, p.10.

**2** Graham Sheffield and Kate Bush, *Martian Museum of Terrestrial Art*, op. cit, p.8.

**3** Gilane Tawadros and John Gill, 'We are the Martians', in Gilane Tawadros, John Gill and Jens Hoffmann, *Alien Nation*, London: ICA and InIVA, 2006, p.1.

**4** Douglas Fogle, 'Is there Life on Mars?', in ed. Michelle Pirano, *Life on Mars: 55th Carnegie International*, Pittsburgh: Carnegie Museum of Art, 2008, p.21.

**5** Ibid. p.27.

**6** Ibid. p.29.

**7** Massimiliano Gioni, catalogue essay, *After Nature*, New York: New Museum, 2008, unpaginated.

**8** Ibid.

**9** Ibid.

**10** Pil and Galia Kollectiv, introduction, *The Institute of Psychoplasmics*, London: The Pump House Gallery, 2008, p.6.

**11** Daniel Birnbaum, 'On Human Nature', in *Life on Mars*, op.cit. p.57.

**12** Ibid.

**13** James Heartfield, *The Death of the Subject, Explained*, Sheffield: Sheffield Hallam University Press, 2002, p.67.

**14** Josie Appleton, 'Recentring Humanity', in ed. Dolan Cummings, *Debating Humanism*, London: Societas Imprint Academic, 2006, p.93.

---

J.J. Charlesworth <jjcharlesworth@artreview.com> is a freelance critic and reviews editor at *Art Review* magazine. He is a member of the Manifesto Club's Artistic Autonomy group

# ORIENTALISM INVERTED: THE RISE OF 'HINDU NATION'

Is Indianness just a German ideology? In the first of a two-part analysis of neoliberalism in the subcontinent, Neil Gray traces the history of Hindu cultural nationalism, from a colonialist mystique of pure spirituality to today's fascist pogroms and economic polarisation

The abject poverty and extreme economic polarisation created by neoliberal regimes requires national ideology to legitimate and obfuscate its violence. Cultural nationalism gives coherence to the activities of the nation state and capital, and best serves the conflicting requirements of accumulation and legitimation for neoliberal elites. Thus Radhika Desai contends:

> ... the deployment of the language of particularity, of cultural difference and nationalism, in counterfeit answer to the accelerating universalism of capitalism, which it supports and promotes, is the ingenious reality of the right today.[1]

In India, the chief cultural nationalist movement is Hindutva, a communalist Hindu nationalist ideology seeking to conflate the very idea of 'Indianness' with 'Hinduness'. The core practitioners of Hindutva are organised under the umbrella of the Sangh Parivar organisation, which is avowedly inspired and influenced by the Rashtriya Swayamsevak Sangh (RSS) a 'social and cultural organisation' with a known fascist pedigree and a Hindu majoritarian political agenda. The importance of this movement, a deeply conservative multi-headed Hydra, can be measured by the presence within its ranks of the former ruling party of India, now the main party of opposition, the Bharitiya Janata Party (BJP). The Vishwa Hindu Parishad (VHP) and the boot boys of the Bajrang Dal, '... the violently energetic youth wing of the VHP'[2] complete the Sangh Parivar 'Trident', which fronts a host of other organisations with Hindu nationalist sympathies.

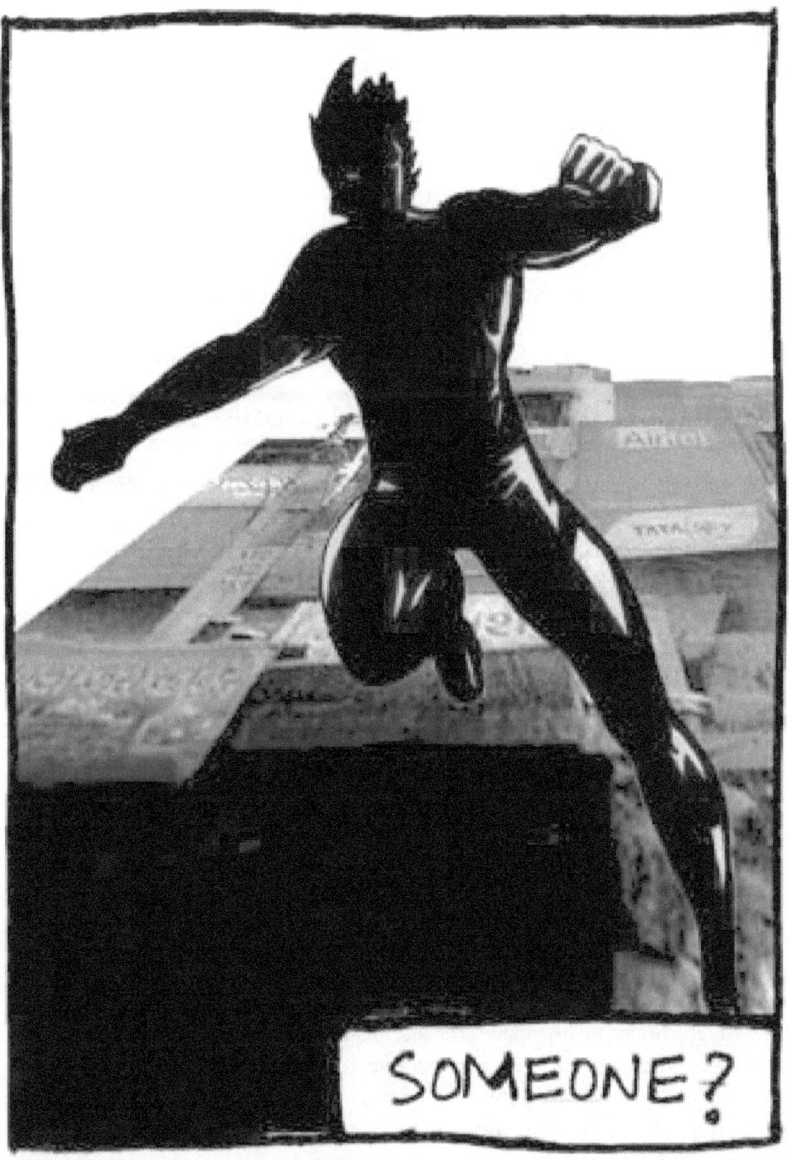

Images: *Raj Comics for the Hard Headed* by Amitabh Kumar (from Sarai Media Lab's Research Project on Raj Comics and Graphic Novel Culture in Delhi)

This book is available for free PDF download from:
http://www.sarai.net/publications/occasional/rajcomics-forthehardheaded

The roots of Hindu cultural nationalism lie, at least in part, in an *inversion* of romanticist orientalist epistemologies of the 19th century. This inversion effectively shifted social and political issues from the material to the spiritual plane – serving the needs of both the colonial masters and the privileged elites of Brahminical Hinduism. Sangh Parivar and Hindutva forces exploit this highly constructed mystical carapace as a counterfeit response to contemporary expropriations under neoliberalism. In this they are facilitated by Western orientalist perceptions of India as an 'essentially' religious civilisation. Augmented by neoconservative theorists like the execrable Samuel Huntington, contemporary Hindu nationalism acts as a neoliberal alibi, masking the extreme authoritarianism and primitive accumulation strategies of international and local capitalist elites in a supposedly 'Shining India'.

## From Orientalism to Hindu Nationalism

> That is the thing about secrets, and that is why we are so eager to know them. They give us, once revealed, a false impression of wider knowledge.
> – Sukheta Mehta, *Maximum City*[3]

Orientalism, as defined by the chief theorist of the term, Edward Said, can be critically discussed as

> ... the corporate institution for dealing with the Orient – dealing with it by making statements about it, authorising views of it, describing it, by teaching it, settling it, ruling over it.[4]

In short, orientalism can be seen as an enormously systematic and diffuse discipline for dominating, managing and *producing* the Orient. Thomas Blom Hansen, following Said's epistemology, has observed that notions of Hinduism as a unified religion, the concept of 'Hindu' as a well bounded cultural category, and Hindu culture as a distinct cultural zone, are largely products of

> ... interventions by orientalist scholars, missionaries, and colonial administrators in the Indian subcontinent since the seventeenth century.[5]

In line with dominant Western epistemologies of the time, early scholars and missionaries attempted to abstract an intelligible core of central tenets and 'scriptures' from their encounter with the vast corpus of religious images, myths and practices in the subcontinent. The resultant 'identification and construction' of classical Hinduism, now conceived as a *unified* religious civilisation, was borne from the supposition of a common Aryan or Brahminical higher caste culture, said to be knit together by a common language – Sanskrit. This common culture was allegedly bound by a body of ancient texts, and sanctified by a sacred geography inscribed by centres of pilgrimage all over the subcontinent. This conception of orthodox or

## the works of German orientalists produced idealist, romanticist readings of India

'classical' Hinduism, however, was in reality codified and constructed from a myriad of diverse and heterogeneous relations of ritual and social religious hierarchies by orientalists such as Max Muller. Indeed, Muller's translation of the *Rig Veda* was seen by the British colonial elites – who commissioned the work – as

> the oldest and thus the most authentic self-born and founding text in the larger body of Hindu philosophy.[6]

This translation, included in his 50-volume compendium *The Sacred Books of the East*, became hugely influential in disseminating the 'Idea of India' as predominantly 'religious', 'mystical', 'spiritual', in both Europe and India.

Hansen argues that the orientalist construction of classical Hinduism resulted in a pre-political 'empty signifier' of national unity: a constructed, essentialised supposition of a national inner life and spirit from which claims could be made on the perennial nature of the Indian people. In the elite hands of orientalists, communitarians and cultural nationalists, this paradigm soon became 'a truth beyond representation and falsification', and

> a metaphysical construct of what should be there in order to make the other intelligible within a system of systematic differences.[7]

It was now possible to identify the East from the West in a single conceptual grammar of civilisational order and hierarchy. Presupposing the apolitical, spiritual character of the Indian masses gave narcissistic license for the colonial and Brahmin elites to lead a *moral* revolution of the people, while separating the imagination of society, state and economy from the profanity of ordinary politics by ordinary people. The orientalist codification of upper caste Brahminical practices into the central tenets of Hinduism, or in the historian Romila Thapar's term, 'Syndicated

Hinduism', was thus central to consolidating the British Empire's hegemonic rule in India, yet it would also instigate the genesis of Hindu nationalism.

## Inverting Orientalism

The colonial government's need for a pragmatic incorporation of elite segments of the Indian middle class led to an indirect, though partial, agency for client groups such as the *zamindari* (agricultural) landlords, literate elites, and leaders of sects, petty kingdoms, and religious communities. These groups, the indigenous 'pillars of colonial rule', were granted some license in cultural matters, if not matters of governance. While property, security and taxation were a preserve zealously protected by the Raj, issues supposedly pertaining to the *Orient* – religion, community and family – were often governed by sanctioned local bodies. Upper caste groups were consequently encouraged by the British to 'know themselves' through the gaze of the other under a high caste religious designation which may not have been so 'rigorously described' previously.[8] According to Hansen, many of these constituencies interiorised the oriental construction of East and West as essentially different, but rather than retain the negative connotations of difference inevitably produced by the colonial regime, they *inverted* the terms

> and reversed the valuation so that the differentiation became a source of cultural and moral superiority.[9]

The empty signifier produced by the British helped Brahmins and other upper caste groups to codify existing social and ritual hierarchies, while consolidating their dominant social position as arbiters of truth and social sanction. The Cultural Nationalism of late 19th century India thus grew from a set of processes whereby influential Indians

> began to inhabit, and make sense of, received romanticist notions of authenticity and deep cultural differences between East and West.[10]

Thus Swami Dayananda of the late 19th century socio-religious Arya Samaj organisation could concoct an influential mythic 'Vedic Golden Age': an empty signifier par excellence, and a myth of which so little was known 'that all fictions could be accepted as valid'.[11] Meanwhile, R.S. Golwalker, the fascist head of the RSS organisation for over 30 years, drew upon the 'founding myth' of the Golden age, to make the erroneous claim that India was first and foremost a Hindu Nation:

Neil Gray

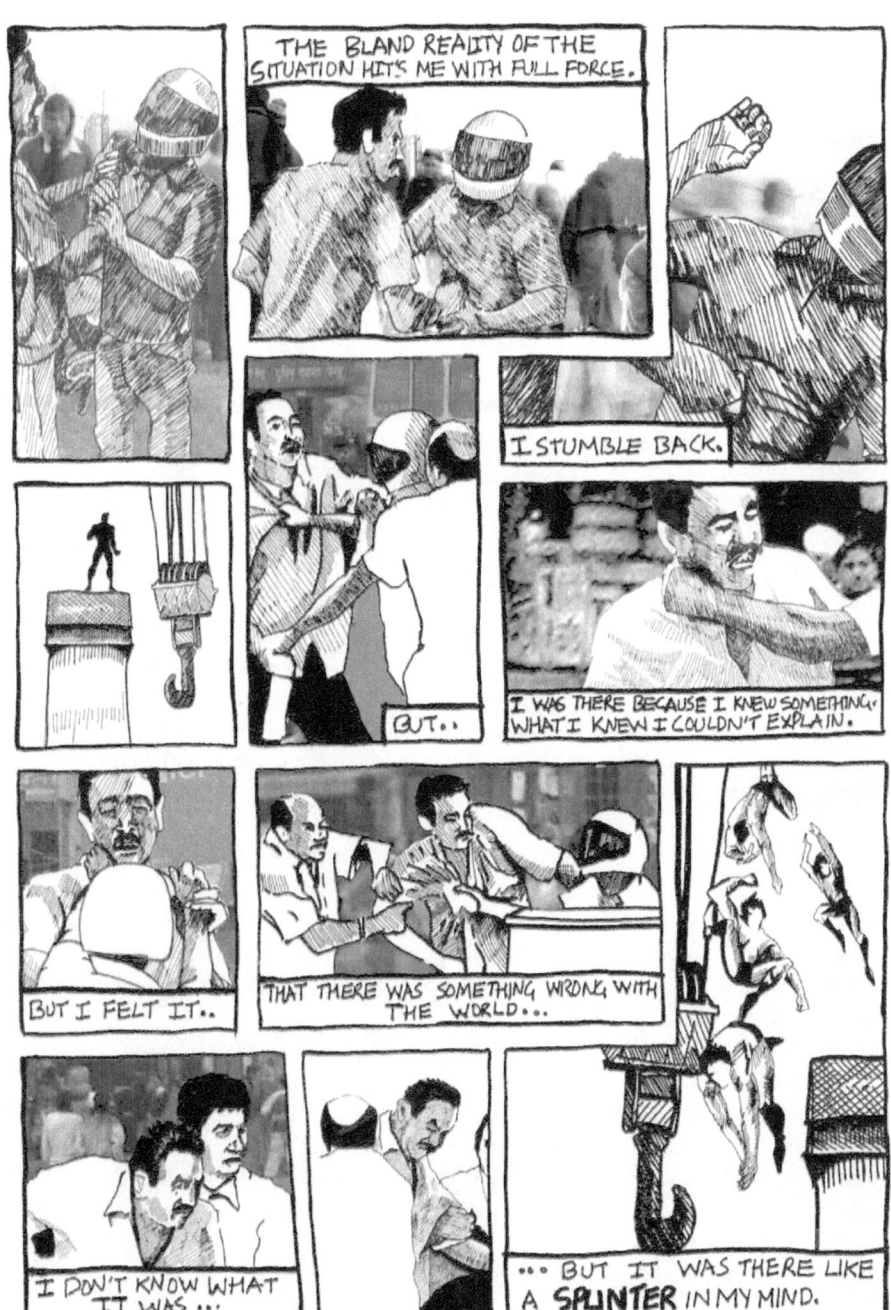

here was already a full-fledged ancient nation of the Hindus, and the various communities which were living in the country were either here as *guests*, the Jews and Parsis, or as *invaders*, the Muslims and Christians.[12]

## Made in Germany

It is in the Orient that we must search for the highest Romanticism.
– Friedrich Schlegel, 1800.[13]

By 1830 Germany had achieved 'intellectual authority' in terms of specialised European scholarship of the Orient, and it was the Germans who were to refine and elaborate the discrete and specialised techniques of reading and analysing texts, myths, ideas and languages 'literally gathered from the Orient' by the imperial institutions of Britain and France.[14] Among parts of the European public, the works of the German orientalists produced distinctly idealist, romanticist readings of India. The romanticist conception of the subcontinent set India for a role as 'spiritual heroine' in the 19th century European anti-modern critique of utilitarian rationalism and crude universalism, and India was widely constructed in the Western imagination as

> a locus of pure essences, of immobility, of high spirituality, and an embodiment of an organic, unfragmented community.[15]

This conception of India, defined in *moral* opposition to the crude material rationalism of the West, became

> an important repository for radical dreams of pristine existence and the whole and the healed self.[16]

For Said, citing V.G. Kiernan, this was 'Europe's collective day-dream of the Orient', lurking in such places as '... the "Oriental" tale, the mythology of the mysterious East, notions of Asian inscrutability ...'.[17]

In a typically orientalist formulation of the time, Friedrich Schlegel, the German idealist philosopher and linguist, could happily exclaim that 'spiritual holism' was the defining feature of Indian culture. The romanticist attraction to spiritual holism, shared by many 19th century German thinkers, mirrored the apolitical stance of India's 'virtuous men from the upper castes'. Spiritual holism collapsed the spiritual and material world into oneness, while at the same time eradicating the separation of the objective world and individual consciousness through incorporation into 'an all-pervasive spirit'. Syndicated Hinduism, in this paradigm, was confused with India per se, and the complicated heterodox traditions and cultures of India were simply conflated with the communitarian ideal of a unified whole: 'India was Hindu' and 'classical Hinduism' was constructed as the '... epitome of holistic spiritualism'.[18]

According to Amartya Sen, the tendency towards 'heroic generalisations' was always evident in the romanticist view of India. On the one hand, Johann Gottfried von Herder could loftily proclaim that

> Hindus are the gentlest branch of humanity [...] moderation and calm, a soft feeling and a silent depth of soul characterise their work and their pleasure, their morals and mythology, their arts.[19]

On the other hand, Schlegel attacked the European man with venom: '... man himself has almost become a machine [and] cannot sink any deeper'. Schlegel's hyperbole even extended to a guarantee that Persian, Greek, Roman and German languages and cultures '... may all be traced back to the Indian'.[20] For Schlegel, the answer to the alienation and degradation of man under the machine age was not to be found in the conscious organisation of labour against capital, but in the great spiritual traditions of the Indian subcontinent. Hegel too got in on the act, making the wildly orientalist claim that India had '... existed for millennia in the imagination of the Europeans'.[21] He understood India as essentially 'pure spirit', though of an imaginative order (soft and feminine) rather than the 'higher' order of the rational (masculine) spirit of the West. For Hegel, echoing the Raj's political instrumentality, this preponderance of the imaginative over the rational explained the enfeebled, fragmented socio-political culture of the Indian states.[22]

# A Hindu Volk

The grandiose assertions of the German romanticists struck a receptive chord with parts of the Indian intelligentsia. Herder's romanticist nationalist philosophy of a nation *beyond politics* residing in the permanent 'life force' of the people and enunciating popular truth in the face of domination appeared 'eminently meaningful' to large parts of the Indian colonial middle class. No mere 'German Ideology', the idea of nation as popular, cultural and latent, spread rapidly throughout India with cultural nationalism quickly developing as the inverted offspring of German orientalism. For Herder, the national soul was '... the mother of all cultures upon earth', representing an inexpressible spirit in the world, which resided in its 'purest form' within the common national Volk.[23] Herder's romanticist discourse of cultural difference and authenticity provided a conceptual grammar

## Orientalism Inverted

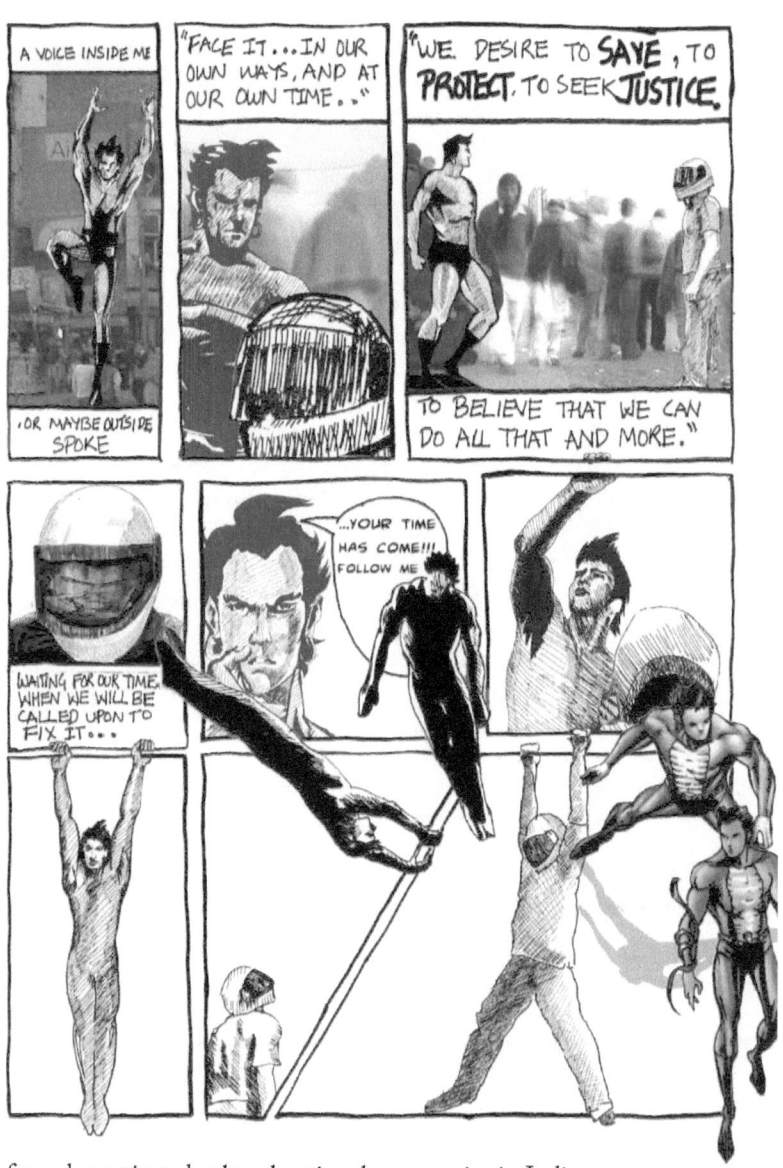

for a domesticated cultural nationalism, and became a powerful impulse for an incipient national ideology based on received orientalist categories in India.

J.G. Fichte further contributed to an essentialised and organicist conception of nation by arguing that cultures were constituted through the nature-given essence of nationality and could only survive and develop through deep emotional attachment to a state '... that gave body to the nation'.[24] By virtue of this profound emotional attachment, a nation could become practically invincible according to Fichte – even in the face of inferiority in terms of material, military and productive power. Cultural nationalism would ultimately depend on 'will' and the 'idea' of nation. The will to sacrifice and loyalty could 'elevate' patriotic men above the petty concerns of politics and historical contingency to provide the very life force of the nation – an idea all too amenable to Indians pinioned by the brute force of colonial domination. The cultural nationalism of Herder and Fichte, and their romanticist emphasis on discourses of '... fullness, spirituality, depth, sensitivity and authenticity', helped ensure that later attempts to construct and consolidate a 'Hindu community' by leading Hindu nationalists would remain captive to the orientalist imagination.[25]

The basis of Vivekananda's 'complete man' was an eternal Herderian national spirit. Like the ideal national citizen of Fichte, he would be educated in 'the great truths of the Vedanta' – culturally awakened and a fervent carrier of the national will.[26] Bal Gangadhar Tilak likewise glorified the deeds of the Vedic age, and in his writings rejected the right of foreigners to criticise or judge Hindu civilisation. Echoing orientalist scholars he claimed the Vedic civilisation was the oldest, the most refined, and the mother of all civilisations in the world. For Tilak, its resilience was proof of its viability. Golwalker, meanwhile, mobilised a series of empty signifiers, in classical orientalist mode, to portray Hindu civilisation as the 'first thought givers to the world', Indian spirituality as essentially 'inexpressible', and Hindu-ness as 'too fine to be defined'.[27] Despite the vagueness of Golwalker's definition of 'Hindu-ness', he argued that those who failed to satisfy its criteria, should 'fall out of the pale of real "National" life'.[28]

## Communalism – from Above

> One must never forget that communalism in India is a latter-day phenomenon which has grown up before our eyes.
> – Jawaharlal Nehru, 1936[29]

The institutions of orientalism emerged alongside the vast expansion of colonial domination around the globe. Above all, communalism developed as 'a weapon of economically and politically reactionary social classes and political forces'. Communalism was an intentional construction on the part of the British colonial elite, and was consciously developed and supported by vested interests for its '... capacity to distort and divert popular struggles'.[30] British divide and rule policy exacerbated and inflamed communal tensions, exploiting existing social differences to divert efforts to create a nationally unified opposition to British colonial rule. Communalism was only one part of this divisive strategy, which set

> region against region, province against province, caste against caste, language against language, reformers against the orthodox, the moderate against the militant, leftist against the rightist [...] and class against class.[31]

The Morley-Minto reforms of 1909, codified and consolidated communalism in India. The reforms established separate, though limited, electorates for Muslim voters. Muslim voters were put into separate constituencies for which only Muslims could stand as candidates, and for which only Muslims could vote. Since voters belonged exclusively to one religion, candidates had no need of appealing to voters of other religious designations. They were thus encouraged to elicit the support of Muslim voters on communal lines. The Montague-Chelmsford reforms of

## Communalism was an intentional construction on the part of the British colonial elite

1919 deepened religious sectarianism by establishing separate Hindu and Muslim mass constituencies. Voters in this system, according to Bipan Chandra, were '... gradually trained to think and vote communally [...] and to express their socio-economic grievances in communal terms'.[32] The colonial propensity for classification intensified these divisions, and the operations of the census from the 1870s onwards further cemented mutually exclusive definitions of ethnic and religious types for the purposes of enumeration and control.[33]

Cementing these divisive procedures, the Raj treated Hindu, Muslim and Sikh communities as separate, with little in common. Official favour was extended to communalists over nationalists; the communal press was endowed with extraordinary freedom in comparison to the nationalist press; and communal leaders were readily accepted as spokespersons for their 'communities'. Nationalist spokespersons, by comparison, were treated as representing only a small minority. These strategies combined to engender client groups – position-seeking middle classes – who were dependent on the colonial elite, at the same time as they created religious and ethnic voting blocks which were previously non-existent.[34] Mushiral Hasan argues, for instance, that the ideological contours of a separate Pakistan were birthed from the 'blurred images' of British ethnic and religious categorisation, as middle class sections of the Muslim community embraced an institutionalised and homogenised conception of religious identity, and began to see themselves: '... in the colonial image of being unified, cohesive and segregated from the Hindus'.[35] Mohammed Al Jinnah's transition from a Congress nationalist 'Ambassador of Hindu-Muslim Unity' in 1906, to his position as the 'leader of Muslim communalism' by the late '20s, and his final arrival as the founder and leader of a separate Pakistan, perfectly illustrates this logic of political communalism.[36]

## From Mother Cow to Ram

The empirical knowledge practices of the colonial state were partly internalised by cultural reform movements who strove to organise and systematise a new and abstract cultural community of Hindus '... within a generalised, supra-local, nationalist discourse and imagination'.[37] Sandria B. Frietag argues that communalist groups emerged in the nexus of relations between the 'public' domains that the imperial nation state claimed for itself and the competing forms of 'private' identity formation that communalist groups staged in the public sphere. As the colonial state increasingly negotiated with Indians on the basis of their religious group identity, it was clear that the rewards went to those '... who invoked only certain kinds of identities'.[38] Drawing sustenance from constructed 'identity slots' and '... authorised by the colonial state', communal groups began to experiment and contest their status more systematically in the public sphere.[39] Frietag argues that the cow protection movement of 1880 – 1920 shows how early public expressions of shared religious precepts '... evolved into larger ideological statements about imagined communities'.[40]

The very ambiguity of meaning in heterogenous, localised worship of 'Mother Cow' allowed activists to link local identities and values to a broader ideological movement. The figure of the sacred cow could '... unite popular and high culture; it could serve reformist and traditional ends; it could reach the heart of townsmen and peasants alike'.[41] Utilising the 'dense metaphor' of the Mother Cow, the movement performed the double function of obscuring intra-caste tensions between Hindu groups contesting ownership of rural land, while underscoring sectarian schisms by setting up various groups outside the shared cultural system – Muslim and Christian beef eaters for instance – as demonised and targeted others. For Frietag, the communalised discursive space that emerged from these constructions of group identity can be traced, from the cow protection movement through to contemporary manifestations, by certain shared characteristics – the evocation of an imagined community; communal group identity before the state; 'dramatic moments', including riots; specialised religious vocabulary; public assertion utilising

# Orientalism Inverted

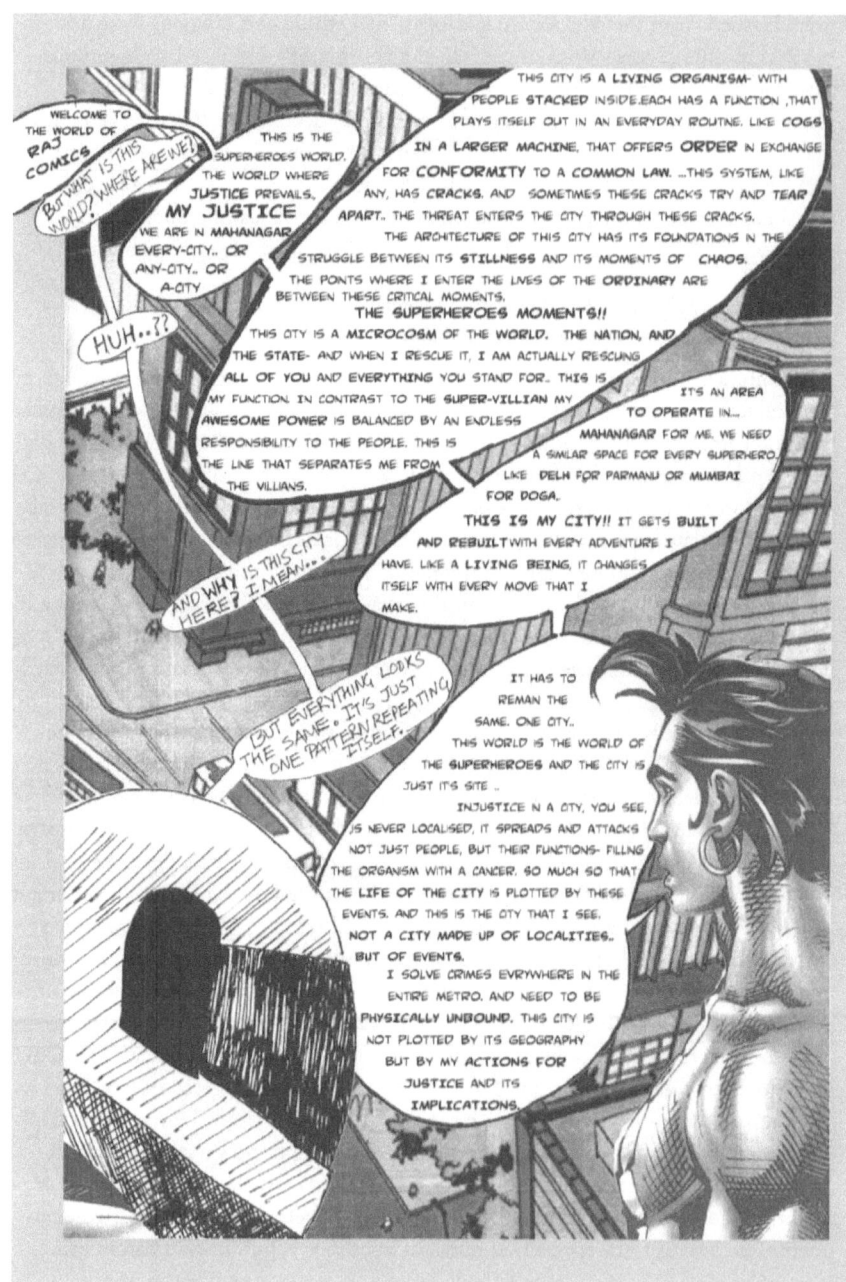

modern mass means of communication; street mobilisation; violence; and the creation of an identifiable 'other', most notably Muslims. As we shall see, all these characteristics, aligned to the mobilisation of the symbol of Ram, were very much in evidence during the infamous events at Ayodhya.

## Ram – 'Prophet' of Hindutva

The colonial assertion of a unified religious basis to Indian civilisation was revived in its inverted perspective during the 1980s through the Sangh Parivar's attempts to turn the worship of Ram, the hero of the oral epic *The Ramayana*, into the paramount God in the Hindu pantheon. Romila Thapar's essay 'A Historical Perspective on the Story of Ram' observes that in Valmiki's 'original' version, and in all the early Bardic tellings, Ram is depicted, not as a god in a religious text, but as a 'human hero' in an oral tale with many local variations. Disregarding this 'all too human' history, RSS ideologists now argue that *The Ramayana* is 'a document of supernatural veracity'[42] and have sought to deploy Ram as 'a metaphor for the catholicity of traditional Hindu forms of devotion and piety'.[43] Ram has now been constructed as a central figure in the VHP's strategy 'to derive, reconstruct and superimpose' an integrated symbolic centre on the large and diverse field of Hindu social practices. The town of Ayodhya in the north of India became the crucible for these strategies, supplying 'materiality and concreteness to the spatial imagination of a Hindu Rashtra [Hindu Nation]'[44].

A temple marking the site of the birthplace of Ram allegedly stood in Ayodhya until its supposed demolition in 1528 by Babur (the founder of the Mughal dynasty). The Babri Masjid mosque, it was argued, was built on the exact site of this purported desecration. For the Hindu Nationalist movement, the site has long held enormous symbolic promise. Plans to mobilise and destroy the mosque have consistently agitated and channeled

> the anti-Muslim sentiment of Hindus towards the mosque as a symbol of their 'humiliating domination' while simultaneously exploiting their feelings of devotion for Ram.[45]

The VHP's decision to reanimate the issue in 1984 was taken despite the fact that there is

> no archaeological evidence to support the idea that a temple ever existed on the site, or that this is the birthplace of Ram, or that the present-day Ayodhya is the site of the capital city of the same name where Rama was born in the Ramayana. *It is entirely a question of belief* [my italics].[46]

Nevertheless, the Ayodhya issue and the adept manipulation of the symbol of Ram helped transform the BJP from a peripheral party of the right into the Congress Party's primary national opposition.

## Ram Versus Barbur – Holy War

In 1990, L.K. Advani, then leader of the BJP, embarked on a national 'Rath Yatra' (chariot procession) designed to awaken in the Hindu population the desire to 'reconstruct' a Hindu temple in honour of Ram on the site of the Babri Masjid. Advani's aim was '... to infuse a sense of shame and humiliation among the people for Hindu society's alleged failure to protect its shrines from desecration by Muslim conquerors'.[47] Ram was openly depicted as an agitational device to embody a virile response to this perceived shame. The Yatra, 'Advani's road show', covered 300 km a day, addressing an average of six meetings daily through eight states in the Hindi heartland. In total, Advani travelled 10 thousand km in a vehicle '... designed to represent an epic chariot and decorated with the electoral symbol of the BJP (a lotus) and the Hindu OM'.[48] Geographical mobility, military display, and provocation all featured heavily. VHP and Bajrang Dal activists lined the way with bows, arrows and trishuls. Advani openly described the mobilisation as '... a controversy between Ram and Babur'[49] while the RSS publicly declared the Rath Yatra 'A Holy War'.[50]

The Chariot procession motif evoked a religious devotionalism to what was essentially a *political* agenda. For the first time a mainstream parliamentary leader unapologetically utilised propaganda '... of an overtly Hindu nationalist character' for political ends.[51] Advani and the BJP projected Ram as a metaphor for '... the essential Hinduness of Indian culture' and the Ram-janmabhoomi agitation was staged as a modern manifestation of '... an ancient, irresistible stream, a corporate Hindu culture'.[52] Conversely, the Babri Masjid was depicted as 'radical negativity' in Hindutva discourses: the embodiment of the '... traumatic historical kernel of Hindu disunity and effeminacy that had to be removed to produce a Hindu Volk'.[53] To this end – against the grain of historic depictions – Ram was reimagined and restaged as a warrior, a bare-chested 'Rambo' with bow and arrows, resplendent in heroic and manly postures. The result of the mobilisation was predictable. The Yatra generated severe communal tensions along its route, '... leaving hundreds of minor and major incidents of anti-Muslim pogroms in its trail'.[54]

In Ayodhya town, on the morning of 6 December 1992, 150 to 200,000 RSS volunteers 'breached' a line of security at the Babri Masjid and began stoning the Mosque and the police guarding it. They used ropes to clamber onto the domes and set about wrecking them with picks and iron rods. The authorities (state and national) abdicated all responsibility to protect the mosque. The media present were attacked and had their equipment smashed. Within two hours in the afternoon all three domes were demolished. Meanwhile, Muslims were attacked in Ayodhya town, and their homes set ablaze. In the week that followed, communal riots officially claimed 1,200 predominantly Muslim lives nationwide. In Bhopal alone, where the riots blazed uninterrupted for a week, 16,895 people were forced to find shelter in 31 refugee camps, while at least seven states were torn by communal riots and badly hit with fatalities.[55]

In the face of evidence from the Archeological Survey of India (ASI), which disputed claims that Ayodhya was the birthplace of Ram, the Sangh Parivar claimed that Hindu matters of faith were outside the rational discourse and 'foreign methods' of scientific investigation. Advani, for instance, asserted: 'No judge can give a verdict on the birthplace of Lord Rama, which is a matter of faith for Hindus'.[56] Meanwhile, Advani was lionised on the BJP website for his role in mobilising Ram during the fiercely communalised events: '... Shri Advani emphasised the cultural unity of the country by highlighting Shri Rama as a symbol of cultural renaissance and as a national symbol'.[57]

## The Ram Sethu Project

The Ram Sethu issue of late 2007 showed that Ram could still be mobilised as a potent force by the Sangh Parivar. The BJP, and groups associated with the VHP, argued that the Sethusamudrum Ship Channel Project (SSCP) – a project to dredge a 167-kilometre passage for container ships between India and Sri Lanka – should not proceed. They argued that any structural change to a group of islands known as 'Ram Sethu' would hurt the religious sentiments of Hindus. In Hindu mythology, Ram is said to have built the Ram Sethu 'bridge' as a chain of islands in order to reach Lanka and rescue his abducted wife Sita. The BJP and the Sangh Parivar argue that belief should be respected in relation to the Ram Sethu project and other large development projects. Further, they argue that 'The Ramayana' constitutes 'historical evidence' of Ram's existence and his construction of the bridge. The Archeological Survey of India's (ASI) affidavit, however, found that the islands were 'natural forms' and that mythological texts could not prove the existence of the characters: '... or the occurrence of the events depicted therein'.[58]

A storm of Hindutva protests over the ASI's alleged affront to Hindu sensibilities caused the Congress-led UPA to withdraw the ASI affidavit, and to redraw the line of the shipping channel for fear of a 'Hindu backlash'. Parful Bidwai argues that the VHP and the RSS now hold 'an effective veto' over the UPA, evidenced most

## Orientalism Inverted

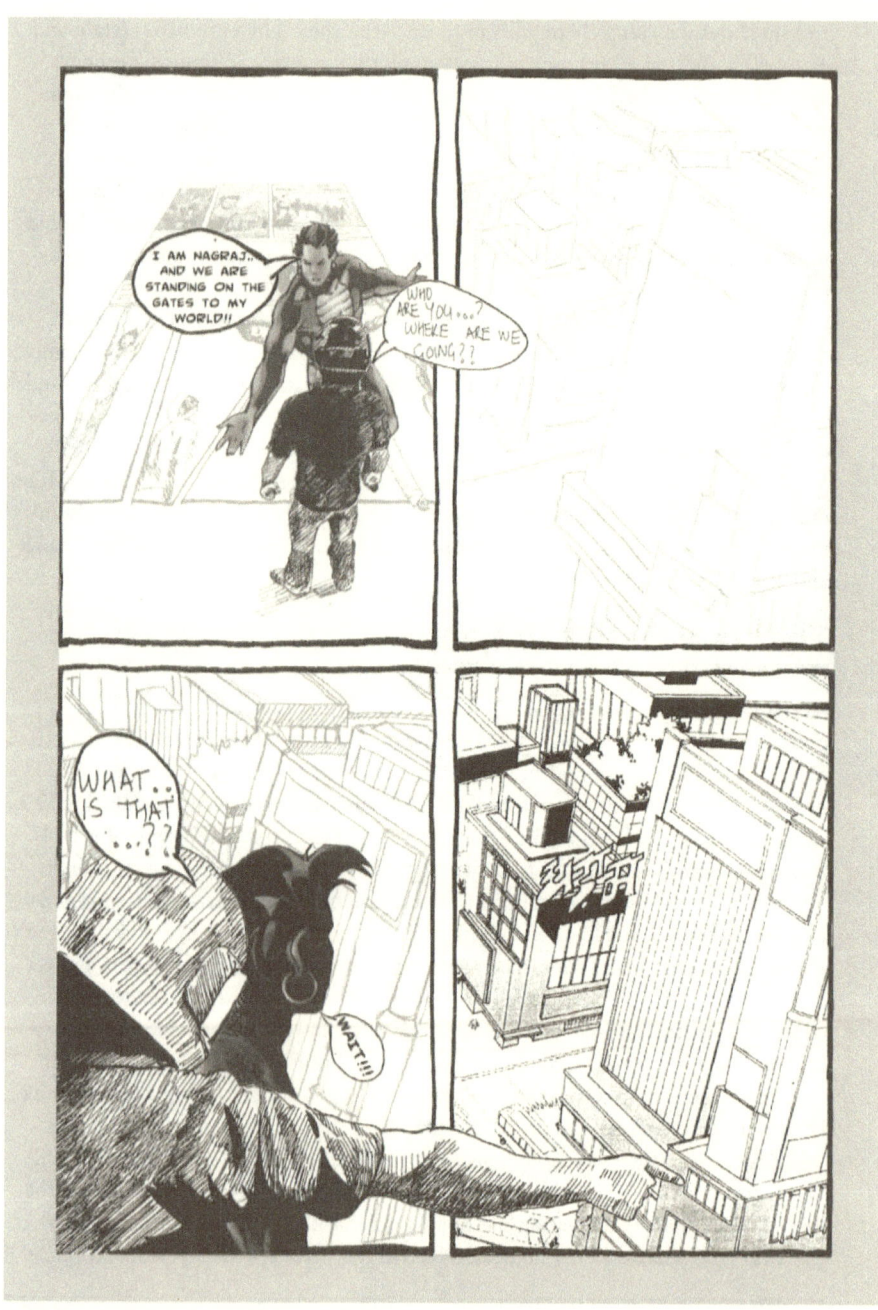

clearly at the political level by the BJP, 'their parliamentary face, alter ego, and ventriloquist'.[59] For the BJP, the ASI's denial of Ram's existence as historical fact constitutes a blasphemous insult to Hindus. L.K. Advani claimed that the 'pseudo-secular' forces of government 'sought to negate all that Hindus consider sacred ... and wounded *the very idea of India*'.[60] The VHP, for their part, warned the UPA government that 'Hindu rage would be unleashed globally against it if it proceeded with the Sethusamudrum project'.[61] The mobilisation of Ram around the Ram Sethu issue was clearly a politically instrumental strategy designed for electoral gain: the Ram Sethu project was first sanctioned by the BJP back in 1998, when the BJP were in power under the leadership of Vajpayee. The turnaround in opinion was designed to agitate 'Hindu consciousness' nationwide for the communally loaded Gujarat elections in the west of India.[62] The withdrawal of the affidavit shows how far the Sangh have come in introducing an irrational religious content, backed with the threat of communalised violence, into national debate, and vividly underscores the surrender of Congress to communal forces. What that has meant in practice was ferociously underlined in the Gujarat massacre of 2002.

## 'Snakes Hissing' – a Hindu Jihad

> We made the whole plan ... to start A Hindu Jehad ... we were successful in Gujarat...
> – Dhimant Bhatt, BJP[63]

The horrific pogrom of over 2 thousand Muslims by Sangh Parivar activists in Gujarat in December 2002 showed that what could not be recuperated through ideological persuasion could just as well be controlled through the judicious use of 'the hard edges of police *lathis* [truncheons]' and the 'dull compulsions of economic relations'.[64] A Human Rights Watch report indicted the Sangh Parivar and the ruling state party, the BJP, for their complicity and culpability in the massacre:

> The groups most directly responsible for violence against Muslims in Gujarat include the Vishwa Hindu Parishad [VHP], the Bajrang Dal, the ruling BJP, and the umbrella organisation, the Rashtriya Swayamsevak Sangh (the RSS).[65]

The report went on to describe Sangh Parivar cadres, as '... militant groups that operate with impunity, and under the patronage of the state'.[66] Desai argues that Hindutva performed a 'major service' to the Gujarati propertied classes by increasing violent competition with the Muslim bourgeoisie of the state. Sectarian ruptures in the new 'religious borders' have now reconfigured Gujarat's urban geography through riots, '... with blatant connections to real estate transactions'. Muslims, who no longer felt safe after the riots, left behind property and position '... to be grabbed by those who feel secure in current conditions'.[67] Despite new evidence of government complicity in the riots – reported in the celebrated *Tehelka* magazine 'sting' – the Gujarat electorate rewarded the BJP by reinstalling the government, and the notorious Narendra Modi as Chief Minister, for their third term in December 2007.[68] Meanwhile, Professor Shastri of the VHP was philosophical about the murderous activity of his 'boys' in the riots:

> We needed to do something. It is said that snakes that are not poisonous should keep the enemy away by hissing once in a while.[69]

The willingness of the Gujarat electorate – one of the wealthiest states in India – to continually re-elect Modi and his BJP government, indicates that the policies and strategies adopted by Hindutva have a wide appeal for the electorally important middle classes. The stabilisation of the Hindutva vote among India's richest, most educated and socially elevated sections has now become fully clear. Hindutva has now proven it features all that this powerful class could wish for:

> Neoliberal economic policies, Hindu cultural assertion, [and] the full range of stances towards Muslims and others with the capacity to disturb their comfortable position by demanding their rights.[70]

## Neoliberal Neo-Ghandianism

> Their innermost secret, namely, that 'there is no secret', only profanity, must, hence, be carefully guarded. It is, after all, precisely this ritualised guarding, this objectification of belief, that generates the illusion that there is a secret in the first place.
> Thomas Blum Hansen, 1999[71]

The brutality of the Sangh Parivar is not in itself capable of creating a hegemonic right in India. For Radhika Desai, neo-Ghandianism is the specific contemporary form of cultural nationalism for the conservative bourgeois Indian intellectual, and is thus central to the restructuring and reproduction of a rightist hegemony in the Asian subcontinent. Neo-Ghandian discourses emerged within the 'intellectual reconstitution' of the global new right in the 1980s and 1990s, which typically conceived of culture as disembodied '... rather than incarnated in society's material – productive and reproductive – processes'.**72** For Desai, neo-Ghandianism, exemplified by Ashis Nandy's writings, borrows from the same irrationalised sources as the Sangh Parivar groups to implicitly, if not always consciously, support Hindutva's quest for hegemony. Desai argues that Ghandi's 'brand equity' provides neo-Ghandians with a claim to tradition and authenticity which allows its followers to do political trade on Ghandi's 'breathtakingly' uncritical international prestige. Moreover, they exploit orientalist ideas about India to deploy mystery and ineffability 'as a product differentiating (secret) patent' in order to assert themselves internationally: '... without having to prove themselves in genuine intellectual engagement'.**73**

Nandy's intellectual coordinates reflect the discourses of neo-conservatism and gain legitimation through bourgeois strains of postmodernism, which carelessly collapse all forms of instrumental and dominating reason into reason itself. Critical traditionalism implicitly conceals caste and class difference, displaces the category of inequality, and reinforces the present hierarchical system '... through a silent abandonment of the analysis of capitalism as a global economic system'.**74**

Critical traditionalism therefore functions as a vital alibi for neoliberalism in India. Nandy has argued that he wants to '... justify and defend the innocence [of the

## Hindutva has a wide appeal for the electorally important middle classes

'non-modern' or 'traditional' colonised cultures] which confronted modern colonialism'.**75** Further, he takes it upon himself to 'speak out' for the 'humble citizen' pitting his body '... against the might of the high technology of his oppression'.**76** By doing so, Nandy reifies Ghandi's conservative discourse of the masses as an empty signifier of political innocence and religious purity – a prime orientalist fantasy of the middle classes.

Regardless of Ghandi's undoubted ability to mobilise the masses against the British, his 'less than emancipatory' role in rural India was pivotal in securing the support of the agrarian propertied classes for the 1920s Congress nationalist campaigns.**77** The Congress left (never hegemonic within the party) was dependent on alliances with rural landlords and rich farmers to deliver the vote banks of the rural poor. They were left 'hostage to Ghandianism' when these landed elites were paid back in kind for their economic and political support. The post-independence constituent assembly secured the economic interests of the rural rich, '... by removing land reform and agricultural taxation from the control of central government'.**78** This decision acted to economically 'constrain the imagination' of the party's progressive left, and bound the Congress, '... to alliances and commitments that came to dominate India's politics until at least the 1960s'.**79** The advance of Hindutva from the '70s onwards was due, in part, to the collapse of this always suspect 'patriarchal-patrimonial' alliance between Congress

# the fantasy of the 'traditional' peasant village persists while its terrain is eradicated

and the rural elites. Furthermore, Ghandi's legitimation of property through his conception of trusteeship, which in Ghandi's words had '... the sanction of philosophy and religion behind it' further catered to the rural dominant classes, as it was finally dependent on religious appeals to morality and the principle of voluntary abnegation of wealth.**80** Ghandi's services to the landed elites can thus be seen as part of his '... flexible ideology *with a record of service to property* [my italics]'.**81**

Nevertheless, Ghandi's 'brand equity' helps Hindutva deploy the fantasy of the 'traditional' Indian peasant village: one without exploited or exploiters; and one where the various class and caste segments are united under the 'chief myth' of the nation as a romanticised and harmonious organic whole. Yet, Nandy's telluric peasant worker, bound harmoniously to the native soil, is more than ever a figment of the orientalist imagination. Hardt and Negri, amongst others, have challenged the romanticisation of the 'traditional' rural worker. For them, we are already in the 'twilight world of the peasant'.**82** Under the globally pervasive ideology of neoliberalism, smallholding subsistence agriculture, the 'natural' terrain of the rural peasant worker, has been almost completely eradicated in the race to achieve economies of scale for export-orientated single crop agriculture. This goal has led to increasing industrialisation, and the consolidation of ownership within agriculture at the expense of peasant smallholders. Thus the picture in Indian agriculture as a whole is 'dominated by a single distinctive phenomena' [sic]; that of the transition in Indian agriculture to increasingly advanced forms of capitalist relations of production,

> with all that it usually means in terms of concentration of productive assets, increasing poverty, and the cultural transformation of production and consumption.**83**

Authoritarian Hindu nationalism, supplemented by pervasive neo-Ghandianism, helps to legitimise the accumulation strategies of neoliberal elites. Resuscitating

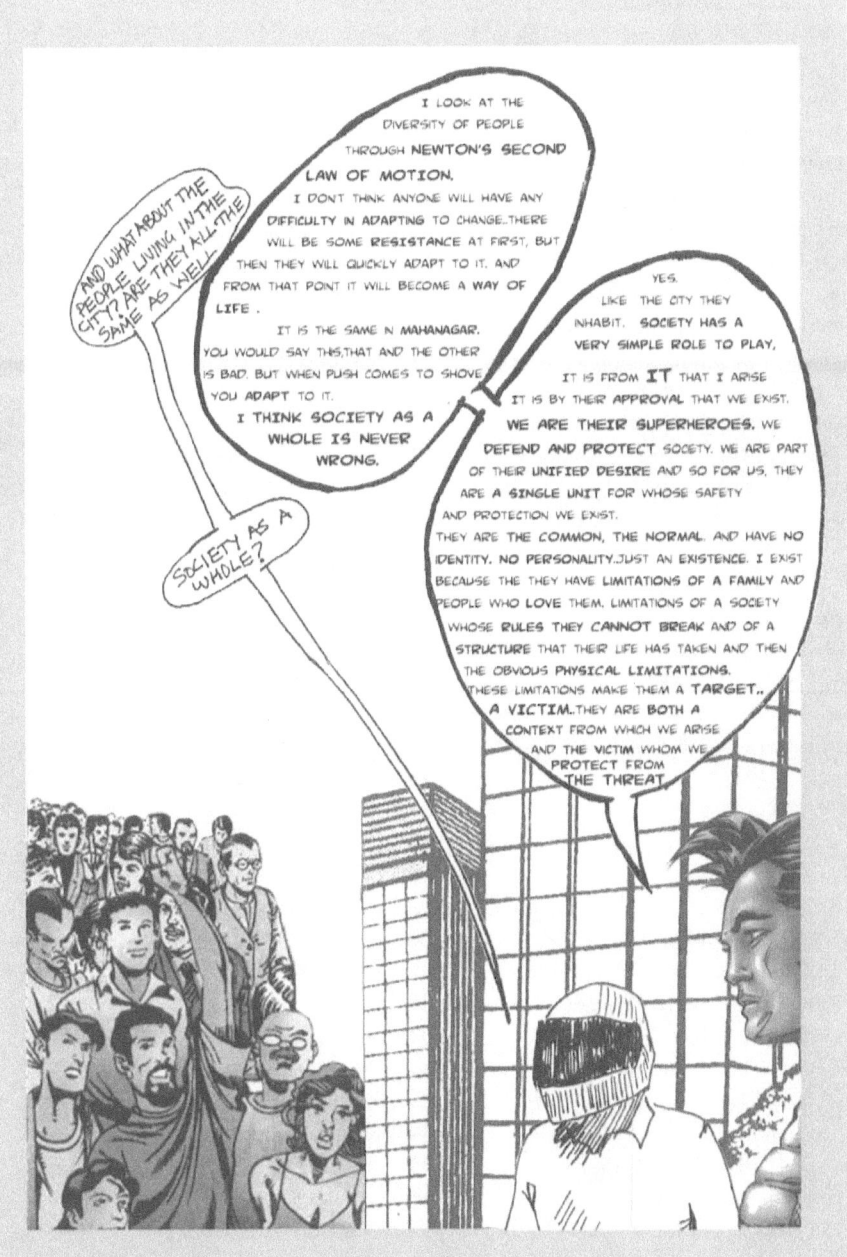

inverted forms of orientalist ideology as a 'counterfeit answer' to the real class and caste content of punitive social and economic regimes, these appear somehow independent from material reality. However, the repetition of 19th century idealism has met with resistance. The next part of this article deals with what Utsa Patnaik termed the 'dangerous classes', those counter forces who ground their theory in the material conditions of production in an attempt to challenge from below what Marx called the 'hegemony of spirit in history'. These recalcitrant forces form the most potent internal threat to the regressive discourses and ideologies of Hindutva.

### Info:

The second part of this article, 'Orientalism Inverted: Resistance in Hindu Nation', can be found at: http://www.metamute.org/en/content/orientalism_inverted_resistance_in_hindu_nation

### Footnotes

**1** Radhika Desai, *Slouching Towards Ayodhya*, Three Essays Press, 2002, p.62.
**2** Amartya Sen, *The Argumentative Indian: Writings on Indian Culture, History and Identity*, London: Penguin Books, 2005, p.52.
**3** Sukheta Mehta, *Maximum City: Bombay Lost and Found*, London: Penguin books, 2005, p.351.
**4** Edward W. Said, *Orientalism: Western Conceptions of the Orient*, London: Penguin Books, 1995, p.3.
**5** Thomas Blum Hansen, *The Saffron Wave: Democracy and Hindu Nationalism in Modern India*, Oxford: India Paperbacks, 2001.
**6** Ibid., p.68.
**7** Ibid., p.66.
**8** Ibid., p.66.
**9** Ibid., p.60.
**10** Ibid., p.42.
**11** Christophe Jaffrelot, *The Hindu Nationalist Movement and Indian Politics*, London: Penguin Books, p.77.
**12** A.G. Noorani, *The RSS and the BJP: A Division of Labour*, Leftword, 2000, p.22.
**13** Edward W. Said, op.cit., p.98.
**14** Ibid., p.19.
**15** Thomas Blum Hansen, op.cit., p.68.
**16** Ibid.
**17** Edward W. Said, op.cit., p.52.
**18** Thomas Blum Hansen, op.cit., p.67.
**19** Amartya Sen, *The Argumentative Indian: Writings on Indian Culture, History and Identity*, London: Penguin Books, 2005, p.152.
**20** Ibid., p.152.
**21** Ibid., p.141.
**22** Thomas Blum Hansen, op.cit., p.68.
**23** Ibid., 41.
**24** Ibid., p.41.
**25** Ibid., p.42.
**26** Ibid., p.70.
**27** Ibid., p.81.
**28** A.G. Noorani, op.cit., p.20.
**29** Bipan Chandra et al, *India's Struggle for Independence*, London: Penguin Books, 1999, p.401.
**30** Ibid., p.407.
**31** Ibid., p.408.
**32** Ibid., p.419.
**33** David Ludden, ed., *Contesting The Nation: Religion, Community, and the Politics of Democracy in India*, University of Pennsylvania Press, 1996, p.279.
**34** Bipan Chandra et al, op.cit., p.409.
**35** David Ludden, ed., op.cit., p.193.
**36** Bipan Chandra et al, op.cit., p.433.
**37** Thomas Blum Hansen, op.cit., p.68.

**38** David Ludden, ed. op.cit., p.219.
**39** Thomas Blum Hansen, op.cit., p.38.
**40** David Ludden, ed., op.cit., p.220.
**41** Ibid., p.218.
**42** Amartya Sen, op.cit., p.xii.
**43** Thomas Blum Hansen, op.cit., p.38.
**44** Ibid., p.161.
**45** Christophe Jaffrelot, op.cit.
**46** Ibid.
**47** K.N. Pannikar, *Communal Threat Secular Challenge*, Earthworm Books.
**48** Christophe Jaffrelot, op.cit.
**49** Ibid.
**50** Ibid.
**51** Ibid.
**52** Thomas Blum Hansen, op.cit., p.174.
**53** Ibid., p.175.
**54** Ibid., p.165.
**55** Ibid.
**56** David Ludden, ed. op.cit., p.54.
**57** http://www.bjp.org/
**58** *Frontline*, Volume 24, Number 19, 5 October, 2007, p.134, http://www.frontline.in
**59** Ibid., p.133.
**60** Ibid., p.133.
**61** Ibid., p.13.
**62** For an extensive account of Hindutva politics in Gujarat, see Neil Gray, *Variant* 32, Summer, 2008, www.variant.org.uk
**63** http://www.tehelka.com/
**64** Ibid., p.102.
**65** http://www.hrw.org/reports/2002/india/
**66** Ibid.
**67** Radhika Desai, op.cit., p.158.
**68** http://www.tehelka.com/
**69** http://www.sacw.net/Gujarat2002/GujCarnage.html
**70** Radhika Desai, op.cit., p.148.
**71** Thomas Blum Hansen, op.cit., p.83.
**72** Radhika Desai, op.cit., p.124.
**73** Ibid., p.72.
**74** Ibid., p.63.
**75** Ibid., p.81.
**76** Ibid., p.82.
**77** Ibid., p.92.
**78** Sunil Khilnani, *The Idea of India*, Penguin Books, 2003, p.75.
**79** Ibid., p.74.
**80** http://www.mkgandhi.org/trusteeship/trusteeship.htm
**81** Radhika Desai, op.cit., p.67.
**82** Michael Hardt and Antonio Negri, *Multitude*, Penguin Books, 2006, p.115.
**83** Radhika Desai, op.cit., p.25.

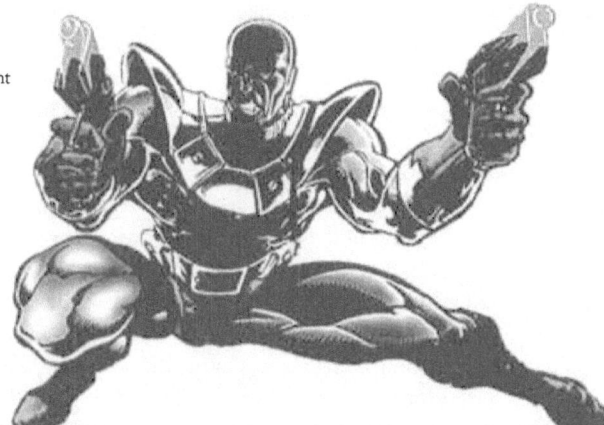

Neil Gray <neilgray00@hotmail.com> is a writer and film-maker based in Glasgow

# ONE WORLD, ONE LIE

The fate of Tibet and its unelected superstar figurehead has captured the attention of Western liberals, not to mention the US government. But the real fascination of Tibet is not its exoticism but its similarity to the rest of an undemocratic global system, argues Paula Cerni

Early this July, the Dalai Lama's envoys returned empty-handed from Beijing. In their first official talks with the Chinese government since protests and riots rocked Tibet in March, the two parties could only agree to meet again in October – once the spotlight of the Olympic Games has dimmed away. The Dalai Lama was disappointed, and so were many Western supporters of Tibetan freedom.

Perhaps, as Slavoj Žižek suggests in a recent article, those supporters are guilty of imprisoning Tibet in their New Age fantasies.[1] 'Our fascination with Tibet makes it into a mythic place upon which we project our dreams', he says. But Tibet's symbolism has less to do with hedonist spirituality than with the harsh realities of life under global capitalism.

Tibet fascinates us not because it is exotic, but because it is becoming more like us, and we are becoming more like it. In just a few years, modernisation has forced upon it contradictions we all recognise. Money and people have poured in, new communication and transport links have opened up, and rates of growth have been high even by Chinese standards. But these changes have also sharpened social inequalities, provoked ethnic conflict, and strengthened the power of elites.

Walk around the streets of Lhasa and you will see shiny new office buildings, shopping malls brimming with goods, and people texting on their mobile phones. You will also see glimpses of the poverty that plagues many Tibetans. But you won't see any independent political organisations, any legal protests or demonstrations against the regime, or any free press. As with the rest of the planet, only more so, capitalism demands political submission while dangling its trinkets in your face.

The most unsettling thing about China, Žižek rightly says, is that her brand of capitalist authoritarianism might be the fate awaiting us all. Could it be, he asks, that 'democracy, as we understand it, is no longer a condition and motor of economic development, but an obstacle?'

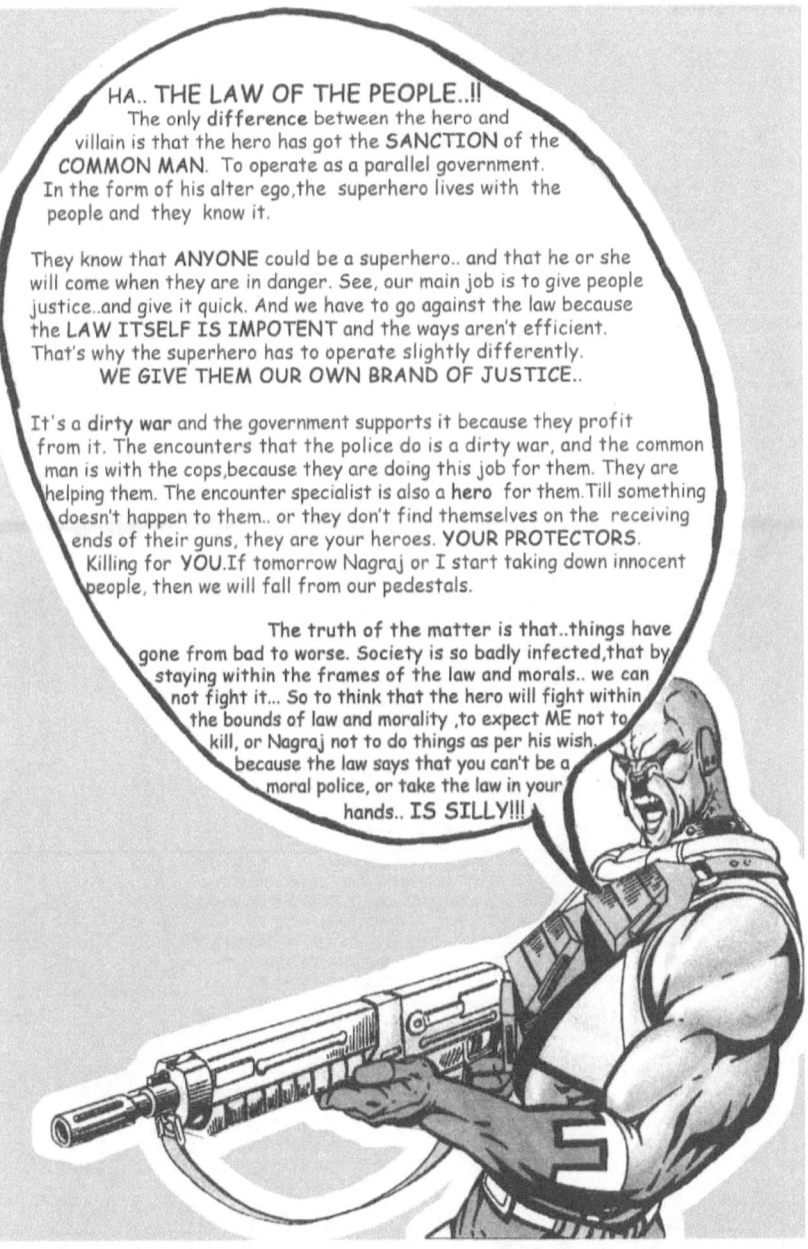

On the contrary, economic development, as we now have it, is a major obstacle to democracy.

It is true, as Žižek argues, that before the PRC invaded in 1949 Tibet was no better, and in many ways was far worse, than it is today. This point has also been

# One World, One Lie

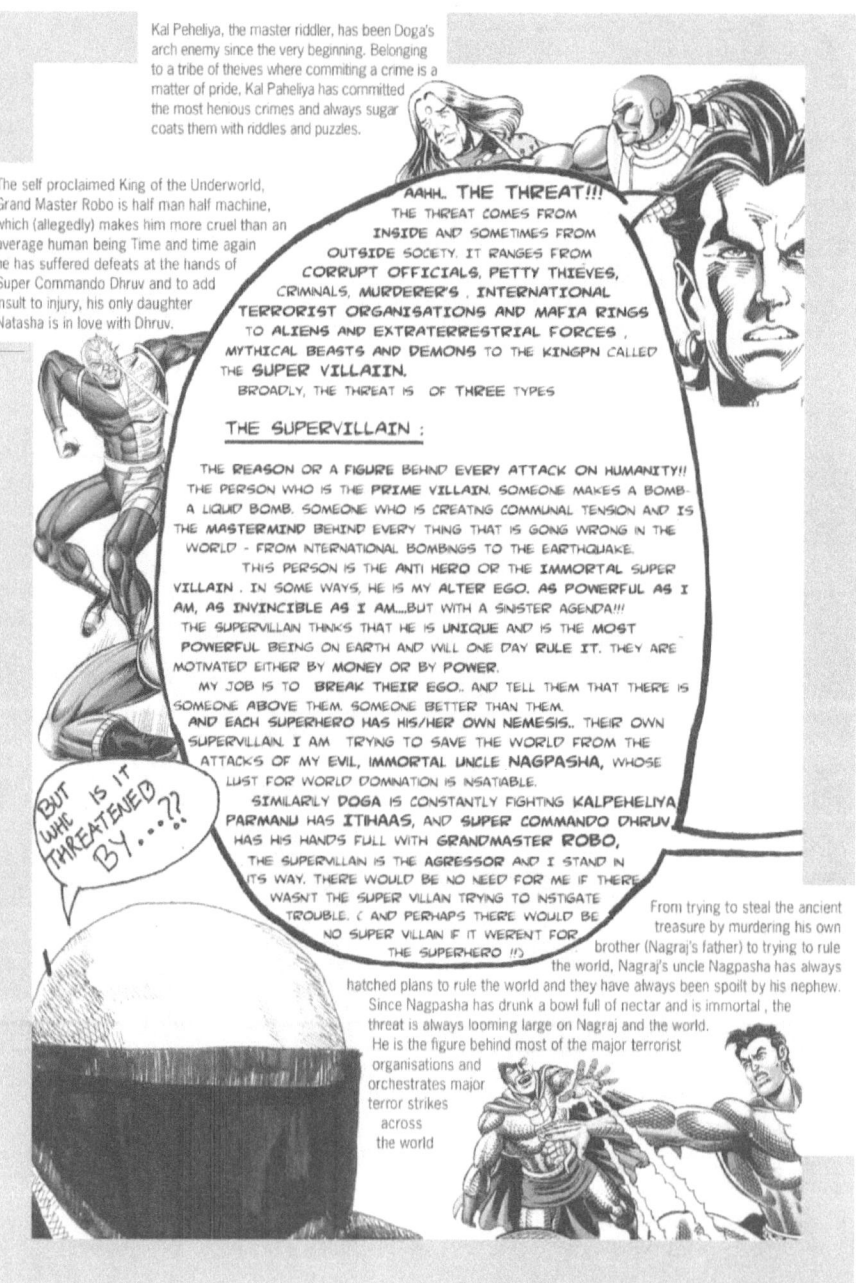

Kal Peheliya, the master riddler, has been Doga's arch enemy since the very beginning. Belonging to a tribe of thieves where committing a crime is a matter of pride, Kal Paheliya has committed the most henious crimes and always sugar coats them with riddles and puzzles.

The self proclaimed King of the Underworld, Grand Master Robo is half man half machine, which (allegedly) makes him more cruel than an average human being Time and time again he has suffered defeats at the hands of Super Commando Dhruv and to add insult to injury, his only daughter Natasha is in love with Dhruv.

**AAHH.. THE THREAT!!!**
THE THREAT COMES FROM INSIDE AND SOMETIMES FROM OUTSIDE SOCIETY. IT RANGES FROM CORRUPT OFFICIALS, PETTY THIEVES, CRIMINALS, MURDERER'S, INTERNATIONAL TERRORIST ORGANISATIONS AND MAFIA RINGS TO ALIENS AND EXTRATERRESTRIAL FORCES, MYTHICAL BEASTS AND DEMONS TO THE KINGPN CALLED THE **SUPER VILLAIIN**.
BROADLY, THE THREAT IS OF THREE TYPES

**THE SUPERVILLAIN :**

THE REASON OR A FIGURE BEHIND EVERY ATTACK ON HUMANITY!! THE PERSON WHO IS THE PRIME VILLAIN. SOMEONE MAKES A BOMB- A LIQUID BOMB. SOMEONE WHO IS CREATING COMMUNAL TENSION AND IS THE MASTERMIND BEHIND EVERY THING THAT IS GOING WRONG IN THE WORLD - FROM INTERNATIONAL BOMBINGS TO THE EARTHQUAKE.
THIS PERSON IS THE ANTI HERO OR THE IMMORTAL SUPER VILLAIN. IN SOME WAYS, HE IS MY ALTER EGO. AS POWERFUL AS I AM, AS INVINCIBLE AS I AM....BUT WITH A SINISTER AGENDA!!!
THE SUPERVILLAIN THINKS THAT HE IS UNIQUE AND IS THE MOST POWERFUL BEING ON EARTH AND WILL ONE DAY RULE IT. THEY ARE MOTIVATED EITHER BY MONEY OR BY POWER.
MY JOB IS TO BREAK THEIR EGO.. AND TELL THEM THAT THERE IS SOMEONE ABOVE THEM. SOMEONE BETTER THAN THEM.
AND EACH SUPERHERO HAS HIS/HER OWN NEMESIS.. THEIR OWN SUPERVILLAN. I AM TRYING TO SAVE THE WORLD FROM THE ATTACKS OF MY EVIL, IMMORTAL UNCLE NAGPASHA, WHOSE LUST FOR WORLD DOMINATION IS INSATIABLE.
SIMILARLY DOGA IS CONSTANTLY FIGHTING KALPEHELIYA, PARMANU HAS ITIHAAS, AND SUPER COMMANDO DHRUV HAS HIS HANDS FULL WITH **GRANDMASTER ROBO**.
THE SUPERVILLAN IS THE AGRESSOR AND I STAND IN ITS WAY. THERE WOULD BE NO NEED FOR ME IF THERE WASN'T THE SUPER VILLAN TRYING TO INSTIGATE TROUBLE. ( AND PERHAPS THERE WOULD BE NO SUPER VILLAN IF IT WEREN'T FOR THE SUPERHERO !!)

BUT IS IT WHO THREATENED BY...??

From trying to steal the ancient treasure by murdering his own brother (Nagraj's father) to trying to rule the world, Nagraj's uncle Nagpasha has always hatched plans to rule the world and they have always been spoilt by his nephew. Since Nagpasha has drunk a bowl full of nectar and is immortal, the threat is always looming large on Nagraj and the world.
He is the figure behind most of the major terrorist organisations and orchestrates major terror strikes across the world

Paula Cerni

made by Michael Parenti in an essay exploding the myth of friendly feudalism.² But feudalism, in Asia as in Europe, never pretended to be so friendly; it was always blatantly about the few ruling over the many.

Only modern societies have combined economic exploitation and political oppression with the promise of popular power. 'We the people' are the first three words of the American Constitution, while its Chinese counterpart declares the people 'masters of the country'. It was a promise extracted as modernity's real economic and social advances thrust the great mass of the exploited and oppressed onto the stage of public life. It was made under duress, and always meant to be broken.

In Tibet, democracy has been smashed to pieces by the same strong currents of economic change that are sweeping through the rest of the planet – by the surge of new wealth spreading from the developing world, especially from Asia, and by the tide of conflict and division coming in its wake.

Tibet's abundant mineral resources can only be thoroughly exploited by Chinese capital if separatism is suppressed. Already, Beijing is investing massively in Tibetan infrastructure in order to ensure easy and efficient access to them. The Qingzang railway, for example, the first to link Tibet with the Chinese heartland, started to operate in 2006. Simultaneously, the Chinese government is building up a layer of social support by encouraging large numbers of Han Chinese – the ethnic group that makes up over 90 percent of China's population – to

## Tibet fascinates us not because it is exotic, but because it is becoming more like us

migrate to Tibet, often on a temporary basis, and favouring them with the best jobs and opportunities.³

Ethnic Tibetans, then, find themselves persecuted and marginalised just as their region's economy is taking off and its subsoil riches are about to be extracted. For them, as for the immigrants who usually hail from poorer parts of China, it is a desperate scramble to survive and succeed, if necessary by trampling over their competitors. During the recent unrest, Tibetan anger at Chinese oppression was mixed with violent physical attacks on immigrants and lootings of Chinese-owned businesses.

There can hardly be a fair settlement unless ethnic discrimination is removed and the population as a whole have religious, cultural, and political freedom – including the freedom to decide whether or not

### One World, One Lie

Tibet should remain part of China. Standing in the way are not only the authorities, but also influential outsiders – the exile organisations represented by the Dalai Lama, and foreign governments.

It is faith in these outside players, as opposed to hope in the capacity of the people to bring about a just and democratic solution, that imprisons many sincere activists, especially in the West, into false fantasies, and, worse still, turns them into unwitting pawns in some very unseemly power games.

I heard the Dalai Lama speak to 50 thousand people at Seattle's Qwest Field Stadium back in April. The sun shone down on a well organised event – while a plane circled above us, pulling a red-lettered banner: 'Dalai Lama, pls stop supporting riots!'

Tenzin Gyatso, the 14th Dalai Lama of Tibet, knew better than to mention the riots. He was not there to discuss international politics with the American public. He asked us to be more compassionate, and reserved his diplomatic moves for a later news conference.

'I am just one human being', he said, his face beaming at us from the giant electronic display screen. But it is not just any human being who claims to be the divine leader of his nation.

The Dalai Lama has never been elected by the people of Tibet to speak on their behalf. His profile is part customary theocracy, part global superstardom, and part leadership of a self-proclaimed government in exile. Of course, it is really the Chinese authorities who prevent everyone in China, not only Tibetans, from electing leaders. Yet the Dalai Lama has brazenly exploited China's undemocratic ways to assert his own right to negotiate with Beijing over the future of Tibet. His constant advocacy of peaceful dialogue, which helped him win the Nobel Peace Prize in 1989, promotes a political process that completely excludes the local population.

Dialogue in this case is the opposite of democracy. Democracy, by placing power in the hands of citizens, would dispense with both the exiled Tibetan monarchy and the autocratic Chinese state. Hence, much as they undermine and frustrate each other, both parties benefit from pinning all hopes on their exclusive talks.

The Dalai Lama is an astute operator who knows which way the wind is blowing. He appreciates the benefits of Chinese-style capitalism and understands that China will never allow Tibetan independence. He is therefore only proposing a 'middle way' of autonomy, though at present even this is too much for Beijing. As he said in broken English during his Seattle news conference,

> A Tibetan should be a citizen of the People's Republic of China. I mean, a happy citizen of the People's Republic of China. I always feel remaining separate, weak, poor. Instead of that, join thousands of millions of people. Prosperity, dignity. Much better.[4]

China, he also told reporters, deserves to be a superpower, but in order to become one she needs moral authority. Since moral authority, as everyone knows, is the Dalai Lama's own area of expertise, this was an offer of help to clean up China's image abroad in return for concessions. Right now, Beijing does not appear to need these services, but perhaps one day it will.

If the Dalai Lama has little real interest in popular democracy, foreign governments have even less, as their policies at home and abroad demonstrate. Still, it feels good to point the finger at regimes more crudely undemocratic than their own, and China's appalling record makes an easy target.

Foreign meddling in Tibetan affairs, however, goes much further than this. The Dalai Lama would never bring it up in front of 50 thousand Americans at a football stadium, but during the Cold War his administration received covert funding from the CIA, much of it allocated to guerrilla operations.[5]

China was then an ideological and political enemy, but still an economic backwater. Today she is the powerhouse of world capitalism. Her competitors are, in equal measure, desperate to do business with her and alarmed at her growing strength. In these circumstances, discretely playing the Tibet card can

## The Dalai Lama is an astute operator who knows which way the wind is blowing

win them 'leverage'. And so we find that many Tibetan organisations and solidarity groups are being aided by powerful foreign sponsors, such as America's National Endowment for Democracy, or the Indian authorities who continue to host the Dalai Lama's shadow government, and tens of thousands of Tibetan exiles, on Indian soil.

Still, in Indo-Chinese relations Tibet is a double-edged sword, plunged into the long standing border dispute between the two nations. While India keeps her Tibetan exiles as a bargaining chip, China takes advantage of the cross-border spread of Tibetan Buddhism to claim sovereignty over Arunachal Pradesh, and particularly its Tawang district, which houses an important monastery.[6]

China is also learning to turn the tables on her other competitors. When pro-Tibetan protesters disrupted the Olympic torch relay in several Western cities – including Paris, London, and San Francisco – Beijing accused them of anti-Chinese racism, and provoked a nationalistic backlash among China's middle classes that included retaliatory protests and calls for boycotts of Western goods. The success of

## One World, One Lie

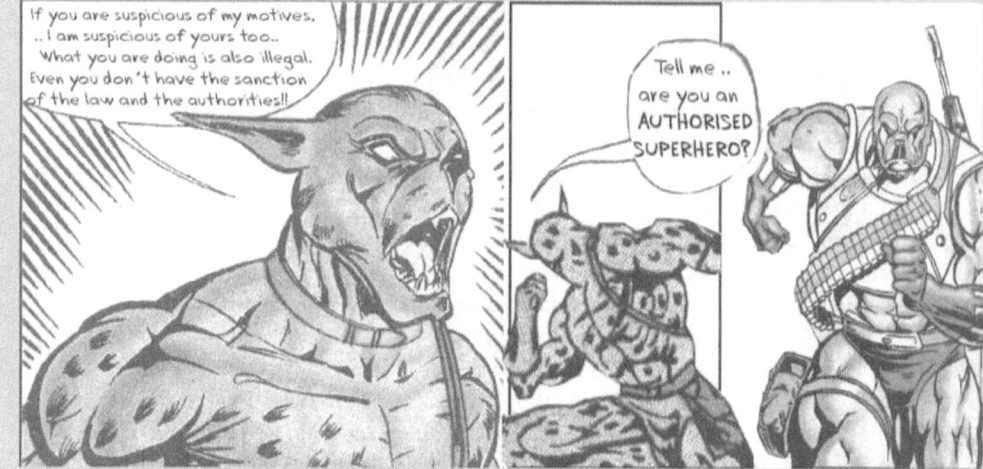

this tactic allowed Beijing to clamp down hard on Tibetans and declare a 'people's war' against separatists, terrorists, and foreign enemies.

China had found her leverage, quickly turning a major challenge to her rule into a political victory. In its aftermath, foreign governments are getting more cautious about their criticisms of China; the Dalai Lama's envoys are getting frostier welcomes in Beijing; and Tibetans are getting crushed.

Standing for the powerlessness that is everyone's reality, Tibet makes the official motto of the Beijing Olympics – 'one world, one dream' – finally come true.

### Footnotes

[1] Slavoj Žižek, 'Tibet: Dream and Reality', *Le Monde Diplomatique*, May 2008, http://mondediplo.com/2008/05/09tibet

[2] Michael Parenti, 'Friendly Feudalism: The Tibet Myth', January 2007, http://www.michaelparenti.org/Tibet.html

[3] Brad Wong, 'Dalai Lama: China Can Change', *Seattle Post-Intelligencer*, 13 April 2008, http://seattlepi.nwsource.com/local/350085_dalaiya14.html

[4] Jim Yardley, 'Trying to Reshape Tibet, China Sends in the Masses', *New York Times*, 15 September 2003, http://query.nytimes.com/gst/fullpage.html?res=9B01E7DF163AF936A2575AC0A9659C8B63&sec=&spon=&pagewanted=1

[5] 'Dalai Lama group says it got money from CIA', *New York Times*, 2 October 1998, http://query.nytimes.com/gst/fullpage.html?res=9C0CEFD61538F931A35753C1A96E958260

[6] Abanti Bhattacharya, 'India reveals flawed Tibet policy', *Asia Times Online*, 7 December 2007, http://www.atimes.com/atimes/South_Asia/IL07Df01.html

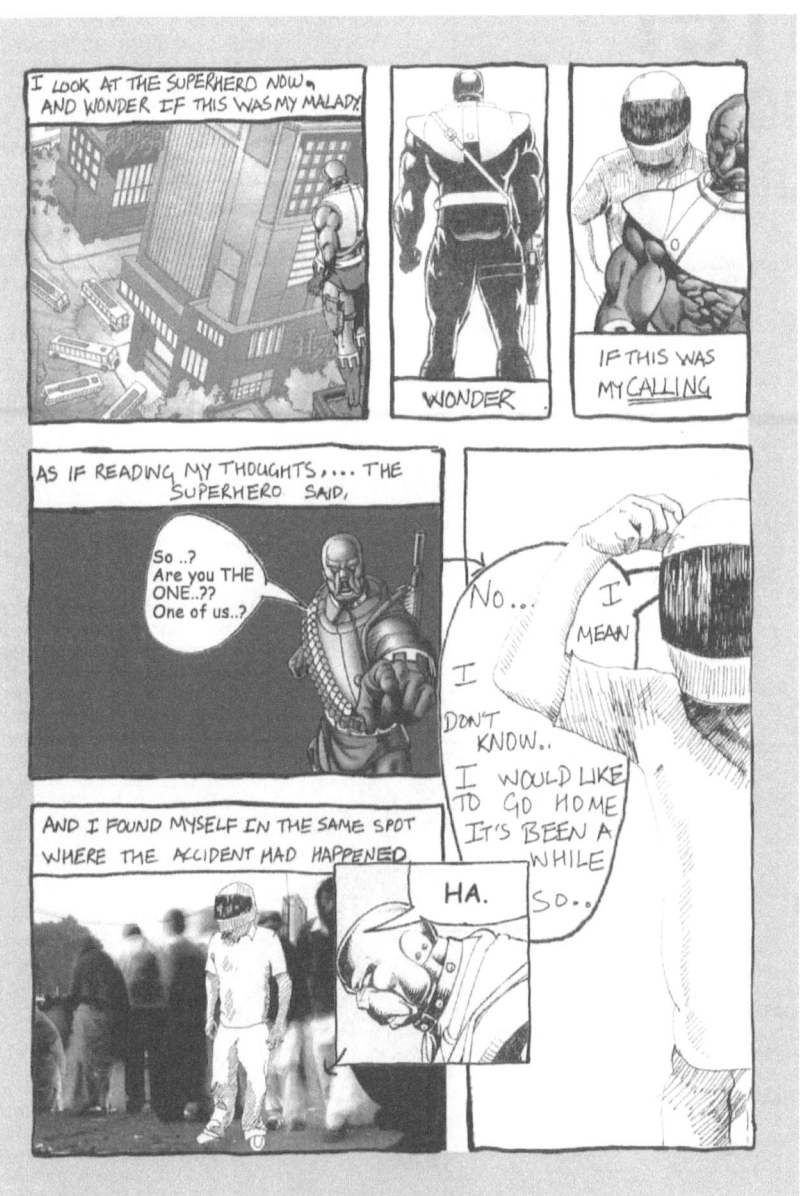

Paula Cerni MPhil is an independent writer. For other publications, please visit http://360.yahoo.com/p.cerni

# Subscription offer

## 10% off the Mute catalogue

Subscribe to Mute and guarantee to be the first in line for our quarterly collection of provocative articles on culture, politics and technology. What's even better, subscribe now and not only get Mute delivered straight to your door but receive 10% off of our new Catalogue, including magazine back issues and titles from the OpenMute print on demand press. That's a year or more of discounts on books, back issues and Mute special projects. Below is just a small selection from the broad range of products on offer.

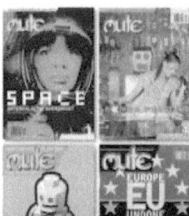

**Find yourself a Mute short of a full set?** Take advantage of our subscriber offer to get 10% off. If you missed out on earlier formats, Mute Back Issues collections offer sets of back issues for only **£35**/collection (which means **£31.50** if you subscribe). Mute's new collections – grouped according to the magazine's successive formats – make it easy to build or complete your very own Mute library*.
- Back Issues I: the Broadsheet (pilot-issue 7, safely packed in two unique pink folders, 'back pack 1 & 2')
- Back Issues II: the Glossies (issues 8-24)
- Back Issues III: Coffee Table (issues 25-29)
- Back Issues IV: the POD (current volume, issues 0-7)

*We regret to say that none of these include issue 9, which is now sold out.

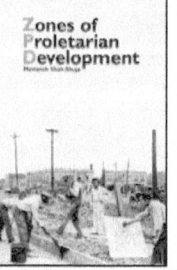

**Shah-Shuja Mastaneh's *Zones of Proletarian Development***
Zones of Proletarian Development is an attempt to theorise the anti-capitalist movement from a neo-Vygotskian perspective. It analyses a series of proletarian activities including recent May Day celebrations in London, carnivalesque football riots in Iran, the anti-poll-tax rebellion and the anti-war movement. Concluding by looking at past and current proletarian organisations, this book makes a number of proposals for future modes of organising conducive to radical consciousness and autonomous activity.
Price ~~£15~~ **£13.50**

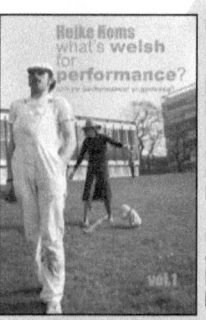

**Heike Roms'** *What's Welsh for Performance? Beth yw 'performance' yn Gymraeg? An Oral History of Performance Art in Wales 1968-2008*
For more than forty years artists have been creating performances, happenings and other time-based art in Wales, yet their work remains largely confined to half-remembered anecdotes, rumours and hearsay. *What's Welsh for Performance?* tries to uncover Wales's hidden history of performance in conversations with key artists who have shaped this history since 1968.
Price ~~£10~~ **£9**

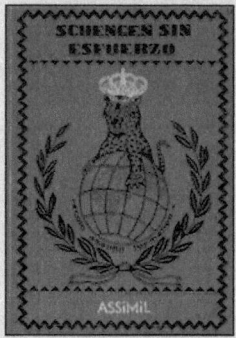

**Vahida Ramujkic's** *Schengen with Ease*
'Extra-comunitarios', or citizens of non-European countries, have the 'extra' bureaucratic task of changing their status to one that will allow them to move and work 'freely' within the European Union. All the required steps are taught through lessons like those found in foreign language skill books, comparing the administrative language of European immigration legislation to an unknown language that has to be mastered first in order to assimilate in to a new environment, receiving determined status.
Price ~~£8.29~~ **£7.46**

Download the complete catalogue at metamute.org/catalogue
Or contact lois@metamute.org +44 (0)20 7377 6949 for a printed copy

# metamute.org/catalogue

# MUTE

## Subscription Rates:

|  | individual | | institutional/company | |
|---|---|---|---|---|
|  | 4 issues (1 year) | 8 issues (2 years) | 4 issues (1 year) | 8 issues (2 years) |
| uk | ☐ £20 | ☐ £38 | ☐ £35 | ☐ £67 |
| europe | ☐ €28 | ☐ €52 | ☐ €48 | ☐ €91 |
| usa/mx | ☐ $40 | ☐ $75 | ☐ $70 | ☐ $133 |
| row | ☐ €34 | ☐ €60 | ☐ €54 | ☐ €102 |

Please tick the appropriate box.

I wish to pay by cheque/credit card.
☐ I enclose a cheque (GBP) made payable to Mute.
☐ Please charge my

☐ Visa  ☐ Access  ☐ Mastercard  ☐ Switch

Card no. ☐☐☐☐ ☐☐☐☐ ☐☐☐☐ ☐☐☐☐
Expiry date ☐☐ / ☐☐
[Switch only] Issue number ☐☐   Start date ☐☐ / ☐☐
Security code ☐☐☐

Signature _____

name _____
address _____
_____
town/city _____
post code _____
country _____
tel _____
email _____

Or call our credit card hotline
Tel +44 (0)20 7377 6949
Fax +44 (0)20 7377 9520

**Online** metamute.org/shop
**Email** mute@metamute.org
**Skype** mute.london

**POST TO: MUTE,
Unit 9, The
Whitechapel Centre,
85 Myrdle St.,
London E1 1HQ, UK**

www.ingramcontent.com/pod-product-compliance
Lightning Source LLC
Chambersburg PA
CBHW020442220526
45464CB00002B/822